SELECTED READINGS IN PHYSICS
*General Editor:* D. TER HAAR

# *The Contributions of Faraday and Maxwell to Electrical Science*

# *The Contributions of Faraday and Maxwell to Electrical Science*

BY

R. A. R. TRICKER, M.A., PH.D.

PERGAMON PRESS

OXFORD · LONDON · EDINBURGH · NEW YORK

TORONTO · SYDNEY · PARIS · BRAUNSCHWEIG

Pergamon Press Ltd., Headington Hill Hall, Oxford
4 & 5 Fitzroy Square, London W.1.

Pergamon Press (Scotland) Ltd., 2 & 3 Teviot Place, Edinburgh 1

Pergamon Press Inc., 44-01 21st Street, Long Island City, New York 11101

Pergamon of Canada, Ltd., 6 Adelaide Street East, Toronto, Ontario

Pergamon Press (Aust.) Pty. Ltd.,
20–22 Margaret Street, Sydney, New South Wales

Pergamon Press S.A.R.L., 24 rue des Écoles, Paris 5e

Vieweg & Sohn GmbH, Burgplatz 1, Braunschweig

---


First edition 1966
Library of Congress Catalog Card No. 66–23859

---

*Printed in Great Britain by Adlard and Son Ltd., Bartholomew Press, Dorking*



(2942/66)

# Contents


PREFACE vii

## PART 1. COMMENTARY

| | | |
|---|---|---|
| *I.* | *Initial Conditions* | 3 |
| *II.* | *The Life of Michael Faraday* | 13 |
| *III.* | *Faraday's Discovery of Electromagnetic Induction* | 24 |
| *IV.* | *The Logical Status of the Law of Electromagnetic Induction* | 41 |
| *V.* | *Faraday as the Founder of Field Theory* | 74 |
| *VI.* | *The Life of James Clerk Maxwell* | 92 |
| *VII.* | *Maxwell's Development of Electromagnetic Theory* | 101 |

## PART 2. EXTRACTS

1. Michael Faraday

| | |
|---|---|
| *Experimental Researches in Electricity* | 135 |
| First Series | 135 |
| Second Series | 154 |
| Ninth Series | 159 |
| Twenty-eighth Series | 169 |
| Twenty-ninth Series | 188 |
| On the Physical Lines of Magnetic Force | 198 |

2. James Clerk Maxwell

| | |
|---|---|
| *Treatise on Electricity and Magnetism* (Chapters VI–VII) | 205 |
| Dynamical Theory of Electromagnetism | 205 |
| Theory of Electric Circuits | 219 |

*A Dynamical Theory of the Electromagnetic Field* (Parts I–IV and VI) 226
Part I. Introductory 226
Part II. On Electromagnetic Induction 238
Part III. General Equations of the Electromagnetic Field 255
Part IV. Mechanical Actions in the Field 268
Part VI. Electromagnetic Theory of Light 276

INDEX 287

# Preface

THERE are two basic principles of electrodynamics. The first describes the action which circuits carrying currents exert upon each other, and its development in the papers of Oersted, Biot, Savart, and Ampère was discussed in a previous volume in this series.* The second deals with the induction of currents. It resulted from work in which Faraday and Maxwell were pre-eminent, and it is with this that the present volume is concerned. As before, the object of the study is to understand something of the genesis of the ideas in this branch of the subject. This is not only capable of adding interest and understanding to the study of electricity, but it also throws light on the logical structure of the subject at the present time.

The growth of field theories is traced to the point at which they left the hands of Maxwell. Starting in a very imprecise way with Faraday, they were given mathematical precision by Maxwell, first by means of a very crude picture of a medium filling space, and later by the application of the methods of generalized mechanics, which made a detailed knowledge of the mechanism of the medium unnecessary. For Maxwell, however, as for Faraday, some mechanical medium remained essential, and in this book we make no attempt to pursue the matter beyond the point where Maxwell left it. The props provided by the medium remain in place.

Unlike the first law of circulation, the necessary foundation for which in modern measurements emerges fairly clearly from a study of the work of Ampère, the correct basis of the second law of circulation—the law of electromagnetic induction—is by no means obvious. The older text-books on electricity, like that of

* *Early Electrodynamics, The First Law of Circulation.*

Jeans, treat it as deducible from the first. The deduction, however, can only be made on the basis of certain other assumptions also, and, with the disappearance of the possibility of a mechanical aether, these have lost their plausibility. Modern writers thus appear to prefer to treat the law of induction as established independently by experiment. Some books, like that of Slater and Frank, go back to Faraday's original experiments for the justification of the premisses on which the theory is built, and others, like that of Panofsky and Phillips, content themselves with a statement that the law of induction is the result of experiment, but do not discuss what the experiments are which are capable of bearing this burden. The original experiments of Faraday, accurate to about one in twenty, although the origin of the ideas, are inadequate for modern requirements, but what is to replace them is by no means clear. The question is of fundamental importance and involves the methods used for the calibration of all voltage-measuring instruments and also the possibility of the measuring mechanical equivalent of heat accurately by an electrical method. The problem is discussed in Chapter IV of the present volume.

Sketches of the biographies of both men have been included since the lives of both possess an interest and stand in such marked contrast to each other. Faraday is the epitome of what can be accomplished by self-tuition and enthusiasm, in spite of the most unpropitious circumstances. Maxwell, on the other hand, had the good fortune to be given an upbringing which was as excellent a training for a physicist as it could well have been. In later life, however, the possession of a beautiful house in the country, surrounded by an estate of six farms which he owned, exerted a considerable temptation to forsake academic life in favour of that of a country gentleman which, in the middle of the nineteenth century, must have been exceedingly attractive and difficult to resist. His wife felt the pull of the country particularly strongly.

The papers by Faraday which are reproduced in Part 2 have been selected to show the essential experimental researches which he undertook. The paper "On the Physical Lines of Force" has also been included as showing the growth of the germ of field theory in his mind. The paper entitled "The Dynamical Theory of the Electromagnetic Field" by Maxwell takes it to the stage of mathematical precision in which it is now used, but it was based upon the possibility of the existence of a mechanical aether, which has disappeared in the meantime. The earlier attempts by Maxwell at a field theory are also discussed in the commentary. They show the beginning of his gradual, though never complete, emancipation from the necessity to assume a mechanical aether, a point which was not finally reached until long after his day.

The author's thanks are due to the Royal Institution for the portrait of Faraday and his wife and permission to publish it, and also to the Cavendish Laboratory, Cambridge, for that of Maxwell and his wife. He is also grateful to Colonel Ferguson, the present owner of Glenlair, for permission to photograph what still remains of Maxwell's house.

The author is also indebted to Dr. D. ter Haar, the general editor of the series, and to Dr. Mary Hesse, for help and suggestions in the preparation of the commentary.

*Cropston,*
*June* 1966

R. A. R. T.

# PART 1

# *Commentary*

CHAPTER I

# Initial Conditions

THIS book deals with the development of electromagnetic theory following the establishment of the basis for the first law of circulation relating to the magnetic fields generated by steady currents, which was provided by the work of Oersted (1777–1851), Biot (1774–1862), Savart (1791–1841), Laplace (1749–1827) and Ampère (1775–1836). Here we shall be concerned with the provision of the basis for the second law of circulation—the law which deals with the induction of currents—which was predominantly the work of three British physicists, Faraday (1791–1867), W. Thomson, later Lord Kelvin (1824–1907) and Maxwell (1831–79), of whom Faraday and Maxwell made by far the larger contribution. Others outside Britain, such as Weber (1804–91), Neumann (1798–1895), Helmholtz (1821–94) and Henry (1799–1878) played a part in it. The subject was opened by the first paper which appears in Faraday's *Experimental Researches in Electricity* and which was communicated to the Royal Society in 1831. From it and other of his experimental researches Faraday developed a number of original ideas which, in the hands of Thomson and Maxwell, led to the field theory of electrodynamics, a development which it will also be the purpose of this book to discuss. These ideas did not, of course, grow up in a vacuum, and the way for them was prepared by earlier work which contributed to the climate of opinion in which such growth was possible. In this chapter we shall glance at the knowledge and ideas, themselves in course of growth and change, which existed at the time and which formed the environment to which the new theories had to be adapted and out of which they emerged.

It will, however, not be possible to afford them more than a superficial consideration.

It is important to realize how labile ideas of the fundamental nature of matter, space and energy were at the time. Did a vacuum exist or was space occupied by tenuous matter in some "fourth state"? A distinction was often drawn between "ponderable" and "imponderable" matter, that is between matter which was subject to the laws of gravitation and that which was not. The electric fluids were commonly thought of as belonging to the latter class, and so was heat or caloric. At an earlier time heat had been looked upon as a mode of motion—for example, by Bacon (1561–1626), Boyle (1627–91), Hooke (1625–1703) and Newton (1642–1727)—but that concept had been invaded by the doctrine of caloric, introduced by a school of chemists in Paris early in the eighteenth century. Caloric was thought of as an elastic fluid, elements of which were mutually repellant but attracted by ordinary matter. Particles of matter, therefore, tended to become surrounded by atmospheres of caloric. This, for example, was thought to account for the latent heat of water, the particles of steam being surrounded by a larger atmosphere of caloric than those of water. Atmospheres of caloric surrounding the particles of gases were also put forward as the basis for the elastic properties of these substances. The caloric theory was seriously eroded by the experiments of Rumford (1753–1814) at the end of the eighteenth century, but it was not finally replaced by the mechanical theory of heat until the middle of the nineteenth century, through the work of Clausius (1822–88), Helmholtz and Kelvin. In the meantime, during the first half of the period in which the theory of electrodynamics was evolving, the situation remained confused.

Throughout the period no better vacuum than the Torricellian could be produced. It was not a good approximation to what would now be regarded as empty space. Through it, of course, light and heat radiation were known to penetrate, and it also conducted the electric spark. The wave theory of light, revived by Young (1773–1829) at the beginning of the nineteenth century—his first paper appeared in the *Philosophical Transactions of the*

*Royal Society* in the year 1800—shared the field with the corpuscular theory of emission. The matter remained undecided until the experiments of Fizeau (1819–96) and Foucault (1819–68) in 1850, in which the velocity of light in an optically dense medium was measured. Throughout the intervening period, however, the wave theory gained in strength through the virtual inability of the corpuscular theory to account for interference and diffraction. It lent support to the view that the vacuum, across which light and electricity could pass, was not "mere empty space".

Little was known about the nature of electricity, whether in the form of static charge or current. Static electricity was generally looked upon as the result of one or two electric fluids, the successors in the theories of the old idea of electric effluvia. The current was fairly generally looked upon as the flow of an electric fluid, though it became agreed that it was impossible to tell whether it was a flow of positive electricity in one direction, a flow of negative electricity in the other, or a combination of both. The latter view tended to become the most popular. It was adopted, for example, by Weber in his analysis of the phenomena of electrodynamics in 1846. Of this view Ampère's theory might be taken as typical. He looked upon the two electric fluids as being in combination with each other in uncharged bodies. The first step in the establishment of an electric current was a polarization of the particles of the compound, each of which contained positive electricity at one end and negative at the other. When the current flowed there followed an interchange of partners between adjacent particles, followed by a reorientation, so that the negative fluid was carried in one direction and the positive in the other. Ampère was much admired by Faraday, many of whose theories were very similar to this view of the electric current. Ampère, however, put forward his theory very tentatively—his main interest lay elsewhere.

Magnetism had developed as an independent science, and its theory was based on the assumption of one or two magnetic fluids, on the lines of electricity. The link between electricity and magnetism had been provided by Oersted's discovery of the magnetic action of the electric current, the account of which was published in July 1820. This was followed immediately by a rapid development of the theory of the mechanical action of steady currents, largely the work of Ampère, whose great paper was printed in the *Mémoires of the Academy of Paris* for 1823, though it did not actually see the light of day until later, on account of the slowness of printing, and during the interval certain additions were made. In it Ampère also put forward the view that magnetism could be accounted for on the basis of molecular currents circulating perpetually round the particles of matter. He also suggested that the aether might consist of the fluid formed by the union of the two electricities.

Clear recognition of the principle of the conservation of energy was not attained until the middle of the nineteenth century. The contact theory of the electric pile continued to find supporters for a long time. The theory amounted to an acceptance of perpetual motion in the case of the imponderable electric fluids, even though in their motion they were performing external work. The theory put forward by Davy (1778–1829) of chemical action gradually gained ground but Faraday still found it necessary to continue the argument in Davy's favour.

This failure to appreciate the general principle of the conservation of energy was caused, very largely, by lack of understanding of the nature of light, heat, magnetism and electricity. By the time that Maxwell was working on the problem of electromagnetic field theory, much of this uncertainty had been cleared away but Faraday worked for most of his time surrounded by this state of confusion.

One of the most important factors to which the origin of the electromagnetic field theory in the mind of Faraday was due, was the plotting of the lines of force in a magnetic field. This goes back a very long time previous to the developments with which we are directly concerned in this book. Peter Peregrinus (Pierre de Maricourt) had plotted the lines on a sphere of lodestone as early as the thirteenth century. For this purpose he had employed a magnetized needle. He was struck by the similarity of the lines he obtained in this way to lines of longitude on the earth, and it was he who introduced the name pole for the extremity of the magnetic axis. The use of iron filings to delineate a magnetic field was also a matter long established. It goes back at least to the seventeenth century. Faraday was familiar with the process and from it he derived a great deal of what, for him, became his fundamental mode of thinking, not only about magnetism but about electricity also.

Thermo-electricity was discovered by Thomas Seebeck (1770–1831) of Berlin, in 1822, in the period during which the magnetic action of steady currents was being investigated. It was put to good use by Georg Simon Ohm (1787–1854) in 1826, when he used it to provide a source of steady electromotive force for the experiments which he performed and which led to the law bearing his name. Polarization and internal resistance in the electric pile and the cells then available, prevented their use for the purpose. Fourier published the mathematical theory of the flow of heat in 1822 under the title of *Théorie analytique de la chaleur*, and Ohm developed his theory of the conduction of electricity in direct analogy with it—an early example of the application of a mathematical theory developed in one field more or less directly to one in which totally different phenomena are involved. The application of gravitational theory to electrostatics and magnetism is, of course, another example of such a process, and it played a prominent part in the development of electromagnetic field theory.

The heating effect of an electric current was well known as a qualitative effect throughout the period with which we shall be dealing. The raising of fine wires of platinum to incandescence was one of the phenomena which had attracted attention in the investigation of the early electric cells. It was not, however, until 1841 that Joule (1818–89) published the results of his quantitative experiments in which he showed that the heat produced by a current in a wire was proportional to the resistance of the wire multiplied by the square of the current, or what is the same thing, to the product of the electric tension and the current. Ohm, who favoured the contact theory of the electric pile, had originally expressed his results in terms of the "electroscopic force" used in that theory. Although this was measured by means of an electroscope working on the electrostatic repulsion between two light objects connected to the pile, it was not defined in terms of work and its relation to the electrostatic potentials of Poisson was not clear. In the case of currents driven by a thermocouple, such as Ohm had employed, electroscopic measurements were only possible in principle and could not very well be carried out in practice. He had realized that the electric current flows throughout the volume of a conductor, unlike an electrostatic charge which resides only on the outside, and he had erroneously assumed that his electroscopic force would be proportional to the density of electric charge at a point in his conductors, just as the temperature at a point in a solid might be considered to be proportional to the amount of heat per unit volume contained in an element surrounding the point.

The work of the great French mathematicians of the eighteenth and early nineteenth centuries, in the development, first of the theory of gravitation and then, by analogy, the theories of electrostatics and magnetism, was based fundamentally on action at a distance. The French, after first accepting Newtonian theory with reluctance, had later turned to it as a model and not only worked gravitational theory out in immense detail but had exploited its general ideas in the spheres of electrostatics and magnetism. It is true that their theories contained the elements of a field theory, in that they were expressed in terms of a potential distributed throughout space, but this was looked upon as the mere summation of the potentials from the individual particles or elements of magnetism or charge, of which the system might be considered to be composed. In electrostatics and magnetism the potentials were simply expressions of the laws of inverse squares of the distance like that which Newton had found for gravitation and which for these two sciences had been discovered by Priestley (1733–1804), Robison (1739–1805) and Coulomb (1738–1806). The term "potential" had, in fact, been introduced precisely to indicate that it did not stand for anything actually supposed to exist in space; it merely served to calculate what would happen if matter, electric charge or magnetic poles were to be placed at the point. In the theory of hydrodynamics developed by Euler (1707–83), a more complete field theory had been worked out. Though not specifically couched in terms of vectors, the idea of a three component point function specified for every point of space was present. In this case, of course, space was occupied by a material fluid. The essence of the later field theories of electricity was that energy and momentum could reside in the field in the absence of matter as ordinarily conceived, or at least in its ordinary form. Such an idea was not contained in the theories of Laplace or of Poisson (1781–1840).

Whether the idea of disembodied energy is capable of being strictly maintained in face of a severe positivist attack is, perhaps, an open question. Those of positivist persuasion can point out that energy is never observable apart from matter. Light rays are sensible only when they fall on matter, and electric and magnetic fields can be observed only by the motion of charged and magnetized bodies or of conductors carrying currents. Radiation can be described equally well by means of retarded potentials, and it is possible to maintain that what is not observable is redundant and better left out of theory. While this may be true, there seems little doubt that a field theory, on the lines which Faraday, Thomson and Maxwell gave to the theory of electromagnetism, was more likely to lead to a theory of radiation than one based solely on action at a distance. The idea of energy and momentum suddenly disappearing and reappearing elsewhere would not be very convenient and would be tolerated only if there were the strongest reasons for abandoning theories of transmission, whatever the difficulty of imagining energy to exist in empty space. Throughout the nineteenth century attempts to meet this difficulty were concentrated upon discovering the properties of an aether which could fill space.

Another writer to whom Faraday refers and who helped to form his ideas of field functions was Boscovitch (1711–87) who wrote in the middle of the eighteenth century. Boscovitch replaced material particles by fixed centres of force, each such centre being connected with every other by means of the forces they exert upon each other. He invented a force function to show the variation of these forces with distance; at great distances it approximated to the law of inverse squares but at smaller distances it alternated between forces of attraction and repulsion, thus providing an explanation not only of gravitation but of cohesion as well.

A subject of investigation that played an important part in the formation of Faraday's ideas was electrolysis. It was the field in which his master, Davy, had been pre-eminent, so that its formative effect upon Faraday's mind is readily understandable. Electrolysis was first commented upon by Nicholson (1753–1816) and Carlisle (1768–1840) on 2 May 1800. They set up the first electric pile in England, and in order to secure good contact at the upper plate had placed a drop of water on it. They noticed an evolution of gas round the wire connected to the pile. They had the initiative to follow this up and proceeded to experiment with wires connected by tubes of water, and when they used wires made of platinum they obtained oxygen and hydrogen. They were thus the first to succeed in decomposing water by means of current electricity. The similar action of "ordinary electricity" from "static" machines in decomposing compounds had been known for some time. The decomposition of solutions of metallic salts was demonstrated in the same year by Cruickshank. At first the main interest in this result lay in its bearing on the identity of electricity obtained from static processes (that is "ordinary electricity" as it was then called) and "galvanic" or current electricity.

Davy used the phenomenon in his theory of the electric pile and made chemical action the source of its energy. However, he included an element of the contact theory among his views. The pile on open circuit acquired its initial charge by contact differences, the whole being like a charged Leyden jar. On completing the external circuit this was discharged. Chemical action then removed the exhausted layer of molecules on the surface of the electrodes so that the action could be maintained. Davy's theory of electrolysis was simply that one of the electrodes possessed the property of attracting metals and hydrogen and of repelling oxygen and acid substances, while the other had the opposite property. The residues left in the electrolyte then attacked neighbouring compound particles so that two chain reactions were set up which were propagated across the electrolyte, one in one direction and the other in the other, thus restoring the situation by freeing the solution of free radicals.

Auguste de la Rive (1801–72), was a great friend of Faraday as was also his father, Professor G. de la Rive, whom he first met in Geneva in the early days when, with Davy, he experimented with circuits composed of different liquids and objected to Davy's supposed attractive powers of the electrodes being the basis of electrolysis. He put forward a theory in which charged particles travelled across the electrolytic solution. For example, he imagined charged hydrogen leaving the positive electrode and travelling across to the negative where it was discharged and liberated. Later Faraday invented his own theory of electrolysis which, fitting in with his general synthesis, confirmed him in his way of looking at electrical phenomena as manifestations of a field existing throughout space, or at least through matter extending throughout space.

As a result of his researches Davy had maintained that chemical affinity was electrical in nature. The same idea was maintained by Berzelius (1779–1848) whose ideas may well have had an important influence upon Faraday although he, however, did not accept them all. Indeed, he tended to be very sceptical of all atomic theories. Berzelius's theory of chemical combination was that it occurred between two constituents, each of which could themselves be simple or complex. The particles of each constituent possessed poles charged with opposite charges of electricity. The pole of one particle attracted the opposite pole of neighbouring particles. The opposite poles of particles which thus came into contact with each other were discharged with the emission of heat or light, leaving the compound particle still as a bipolar arrangement. The charges carried by the two poles of the particle of an element, according to Berzelius, might not be exactly equal. In this way he tried to account for Volta's contact potential differences and for his electromotive series.

The period with which we are concerned in this book is roughly from the year 1831, the date of the publication of Faraday's first paper on electricity, to 1873, the year which saw the publication of Maxwell's *Treatise on Electricity and Magnetism.* The 40-odd years which it spans was one of very rapid development in the concepts of physical science. The outline sketch given above of the state of affairs at the beginning of the period will suffice to indicate the confused situation which existed at the time. By the close of the period the position was very different. One of the questions which is of some interest to pursue is why it was that Maxwell succeeded where Faraday had just failed to arrive at an electromagnetic theory of radiation. The two men had a good deal in common though they differed enormously in their training. In trying to understand the reason for the more complete success of Maxwell it must not be forgotten that he came later on to the scene than Faraday, and thus had the advantage of working in a totally different climate. Ideas, which to Faraday came only after laborious investigations and then not always by any means very clearly, had been clarified in the interval and were thus available to Maxwell. He did not have to contend with the morass of uncertainty which Faraday did so much to clear up and the way was more easily open for the key ideas which Maxwell supplied. The difference in the mathematical equipment of the two men was, of course, immense, and it obviously must have had an enormous influence upon their degree of success, but in considering the effect which this must have had, it is important to remember that they did not have both quite the same difficulties to contend with. Faraday's work was virtually completed by the year 1855, the year from which his biographer, Dr. Bence Jones, Secretary of the Royal Institution, dates the period of his decline. Maxwell's first paper on the electromagnetic field was communicated to the Cambridge Philosophical Society in 1856.

CHAPTER II

# The Life of Michael Faraday

MICHAEL FARADAY, one of the most remarkable characters of British science, owed little to his pedigree except for the genius that made him what he was. He was born on 22 September 1791 at Newington Butts in Surrey, on the outskirts of London. His father was a blacksmith from Clapham, Ingleborough, Yorkshire, and his mother was the daughter of a farmer from the Mallerstang valley to the south of Kirkby Stephen. In 1796 the family moved into London and Michael spent his childhood at a house in Jacobs Wells Mews, off Charles Street, Manchester Square. In 1810 the family moved again—to 18 Weymouth Street, near Portland Place—and in that year his father died. His mother lived until 1838 and in later years was supported entirely by Michael.

In 1840 Michael became errand boy to George Ribau, a bookseller, and one of his main tasks was that of a newspaper round. It may be of interest to present-day newspaper boys to learn that then papers were lent rather than sold, so that each customer required a visit to collect the paper as well as one for delivery at the beginning of the day. In the following year (1805) Faraday was apprenticed and 4 years later his father was able to write "Michael is bookbinder and stationer and is very active at learning his business. . . . He has a good master and mistress and likes his place well. He had a hard time at first going but [now] there is two other boys under him."

Faraday was able to read many of the books he bound and became interested in science. He attended lectures by Mr. Tatum and he also had some lessons in perspective from J. J. Macquerier, who was probably a lodger at the Ribaus's. At Tatum's house he met two other youths, Huxtable and Abbott, and with them he maintained a correspondence that lasted for years, particularly in the case of Benjamin Abbott, who was from a Quaker family, a clerk in the City and rather better educated than Faraday himself. The two formed a kind of mutual improvement society and wrote long essays to each other. They are extremely tedious to read and stilted in style, quite unlike other writing of which Faraday was capable. Only a few years later when touring Europe he wrote very attractive accounts of his travels.

In 1812, at the end of his apprenticeship, he became journeyman bookbinder to a Mr. de la Roche, to whom he took a dislike and he became dissatisfied with his work. It was then that he took the liberty of writing to Sir Humphrey Davy at the Royal Institution, enclosing some notes he had made of some of Sir Humphrey's lectures. Davy was impressed and sent for him, and in March of 1813 Faraday was appointed assistant at the Royal Institution at a salary of 25*s*. a week together with two rooms. For a man of 22, of limited education and entirely self-taught at that, this was by no means a low figure at which to start and indeed, after his childhood which was passed in fairly poor circumstances, Faraday was comfortably off for the rest of his life. He joined the City Philosophical Society, of which he became a very keen member. At the Royal Institution he was engaged with Davy on experiments with nitrogen trichloride and appears to have had an exciting time with the stuff exploding.

It was on 13 October 1813, the Napoleonic war being still at its height, that Faraday set out, with Sir Humphrey and Lady Davy and her maid, on a long continental tour. It lasted 18 months and had been planned to last longer, but the final stages to Greece and Turkey were abandoned. Special passports for the journey were granted by the Emperor and the party crossed from Plymouth to Morlaix, to the north-east of Brest, taking their own carriage with them. It was taken to pieces for shipment and reassembled in France, of which process Faraday gives an amusing account. On 24 October he saw a glow-worm for the first time in his life. The party travelled to Paris where they spent some time. There they met many distinguished French scientists including Ampère, Cuvier, Gay-Lussac, Humboldt, Clément, Desormes and also Courtois who had just discovered iodine, then the topic of the day. Davy performed a number of experiments on it both in Paris and, later on the tour, in Italy. His success in solving the problems presented by the new element tended to incense the French chemists who felt that Davy had intruded into their province.

Faraday had previously never been more than 12 miles from London. This continental journey, which brought him into contact with the most distinguished scientific minds in Europe, supplied for him what a university did for others. The tour broadened his intellectual outlook, it gave him a number of foreign friends with whom he later maintained correspondence, and he learnt both French and Italian. In Paris he caught a glimpse of Napoleon, and from there they travelled south. At Montpelier he saw the Pope processing back to Rome, but in general he took little interest in politics.

From Nice their route crossed into Italy via the Col de Tende, which rises to a height of about 4300 ft. It was crossed on the 19th February 1814. The carriage was again taken to pieces and transported over the mountain on sledges, each pulled by twenty men. Mules carried the baggage while Faraday ascended on foot carrying a mercury barometer. Faraday records the temperature on the summit as 11°F and atmospheric pressure as 25·3 inches of mercury. They reached a village on the Italian side at 7 p.m. guided by lanterns and there they passed the night.

March was spent in the north of Italy. In Florence they saw Galileo's first telescope, but they devoted most of their energies to the burning of diamonds in oxygen. In Genoa they experimented, rather unsuccessfully, with electric eels, and from the water front Faraday witnessed three water-spouts out at sea. From Genoa they had an exciting voyage in an open boat in rough weather and reached Rome on 7 April and Naples on 13 May. They made two ascents of Vesuvius, the second being in the evening to view the volcano at night. On that occasion they had "a species of dinner" on the mountain. The menu consisted of bread, chickens, turkey, cheese, wine, water and eggs roasted in the lava on the spot. At its conclusion "Old England" was toasted, they sang *God Save the King* and *Rule Britannia* and two Russian songs; a good time seems to have been had by all.

The party then went north again. In Milan they met Volta, "an hale elderly man bearing the red ribbon and very free in conversation". They crossed into Switzerland and at Geneva they spent some time with Professor G. de la Rive. There an incident occurred which was indicative of some of the difficulty which Faraday experienced on the tour. Professor de la Rive appears to have been impressed by him and asked him to dine with himself and Davy. Davy protested since Faraday was, in some respects, his servant, and de la Rive, after expressing himself very forcibly, finally had Faraday's dinner served for him in a separate room. Faraday became a lifelong friend of both the Professor and his son Auguste. From Geneva they then toured Switzerland and south Germany (Lausanne, Vevey, Berne, Zurich, Schaffhausen, Munich) returning to winter in Rome in December 1814. The carnival appears to have been the principal centre of interest. In the spring they went back to Naples, and on 7 March 1815 Faraday laconically records "I heard the news that Bonaparte was again at liberty. Being no politician I did not trouble myself much about it, though I suppose it will have a strong influence on the affairs of Europe."

Shortly afterwards the tour ended somewhat abruptly. Sir Humphrey had intended to go from Italy into Greece, the Greek islands and thence on to Turkey, but this idea was dropped and the party returned home. They were in Rome on 23 March and Brussels on 16 April, and a week later they were back in England.

Faraday's ambition had always been to enter the field of science. He speaks of "trade which I hated and science which I loved" and of "the glorious opportunity I enjoy of improving in the knowledge of chemistry and the sciences with Sir Humphrey Davy". The tour, however, had not been without its difficulties. At the beginning Sir Humphrey's man had declined to go with him at the last moment before setting out, and Faraday had agreed to fill the gap temporarily until another servant could be obtained *en route*. None, however, could be found, and in the end Davy gave up trying to find one. Faraday was very sensitive to what he felt to be a change for the worse in his status, and this addition to his duties worried him a good deal. Sir Humphrey, he said,

> is always as careful as possible to keep those things from me which he knows would be disagreeable. But Lady Davy is of another humour. She likes to show her authority and at first I found her extremely earnest in mortifying me. This occasioned quarrels between us at each of which I gained ground and she lost it . . . and after each she behaved in a milder manner.

On his return Faraday was re-engaged as assistant at the Royal Institution at 30*s*. a week and this was later increased to £100 per year, house, coals and candles. He gave his first lecture to the City Philosophical Society on 17 January 1816. It was entitled "On the General Properties of Matter". By 1818 his correspondence with Benjamin Abbott petered out and a new correspondence with the de la Rives began. The latter started with the father and continued with the son and was to last altogether for the best part of 50 years.

Faraday was fond of walking, and in 1819 he went for a walking tour in Wales. In that year he also gave a lecture to the City Philosophical Society which is worth mentioning. It was on "The States of Matter" and for it his notes read:

> States of matter: solid, liquid, gaseous, radiant.
> Radiant state: purely hypothetical. Distinction.
> Reasons for belief in its existence: experimental evidence.
> Kinds of radiant matter admitted.

He kept his papers very methodically and included among them was a commonplace book. Among the entries were sundry notes and verses against love. He had to live these down when he became engaged in 1820 to Sarah Barnard, daughter of an elder of the Sandemanian Church, in which Faraday was brought up. Her brother saw the book and told his sister about the entries. She left London for Ramsgate without having accepted Faraday's proposal and he pursued her there. He seems to have been well received for they went out together and visited a number of neighbouring places. Faraday records one specially glorious day they spent on the hills. They were married on 12 June 1821.

In July of the same year Faraday formally joined the Sandemanian Church. This was a small and strict body of what would now be looked upon as fundamentalist belief. Religious devotions started at 11 on Sunday mornings and went on until 1. The midday meal was then taken together and afterwards the proceedings were resumed and continued until 5. Faraday later became an elder of the church and preached on alternate Sundays. Bence-Jones says:

> It is very difficult to draw a comparison between his preaching and his lecturing: first because they were very unequally known, and secondly because of the entire separation he made between the subjects of religion and science. Generally, perhaps, it might be said that no one could lecture like Faraday, but that many might preach with more effect.

Being an elder of the church required attendance at every service, and it is said that Faraday was removed from office for absence on one occasion without adequate reason. His excuse that he had been commanded to dine with the Queen at Windsor was not accepted.

Faraday's early researches were not carried on without difficulties of other than a scientific nature. He became the first to succeed in making an electrical current revolve continuously round the pole of a magnet—the first example of continuous movement produced electrically. This led to a squabble over priorities. Faraday had been present when Wollaston and Davy had discussed the question together and performed an unsuccessful experiment. It appears, however, that what Wollaston expected was to see the wire rotate on its own axis. Later there was another dispute over priorities, this time with Davy about the liquefaction of gases. Davy claimed that he had reasoned that liquefaction might take place prior to the experiments which he suggested that Faraday should make. This took place at the time when Faraday's name was before the Royal Society for election as Fellow. Davy certainly opposed the election actively. In 1835 Faraday wrote:

> I was by no means in the same relation as to scientific communication with Sir Humphrey Davy after I became a Fellow of the Royal Society as before that period; but whenever I ventured to follow in the path

*Michael Faraday and his wife*

Taken from a daguerrotype which belonged to Faraday's niece, Miss Barnard, and kindly lent for reproduction by the Royal Institution.

The Grace and Favour Residence in Hampton Court, the home of Michael Faraday towards the end of his life and where he died in 1867, as it is today.

which Sir Humphrey Davy has trod, I have done so with respect and with the highest admiration of his talents.†

Faraday's work on the liquefaction of gases led to several explosions. He wrote to his friend Huxtable on 25 March 1823:

> I met with another explosion on Saturday evening, which has again laid up my eyes. It was from one of my tubes and was so powerful as to drive the pieces of glass like pistol shot through a window. However, I am getting better and expect to see as well as ever in a few days. My eyes were filled with glass at first.

On the formation of the Athenaeum Club, Faraday was appointed its first secretary. He kept the post for only a short time as it did not appeal to him. He resigned in May 1824.

On 7 February 1825 Faraday was appointed Director of the Laboratory at the Royal Institution, under the superintendence of the Professor of Chemistry. He was instrumental in starting evening meetings from which the Friday evening discourses originated. Later in the same year he undertook experiments on glass with the aim of improving optical instruments. The experiments went on for some years but produced no major development in optical glass, though a chance specimen was instrumental at a later period in leading to the discovery of diamagnetism, a phenomenon which greatly strengthened Faraday's ideas of lines of force and field theory.

In 1829 he was appointed lecturer in chemistry at the Royal Military Academy at Woolwich. He was not indifferent to monetary reward, as the following extracts from a letter he wrote to Colonel Drummond about the appointment show.

> Sir,
>
> In reply to your letter of the 26th and as a result of our conversation on Saturday, I beg to state that I should be happy to undertake the duty of lecturing in chemistry to the gentlemen cadets of Woolwich, provided that the time I should have to take for that purpose from professional business at home were remunerated by the salary. . . . For lectures which I deliver in this Institution, where I have the advantage of being on the spot, of possessing a perfect laboratory with an assistant in constant occupation, and of having the command of an instrument maker and his men, I receive independent of my salary as an officer of the establishment £8 15. 0 per lecture. The only lectures I have given out of this house

† Bence Jones, vol. I, p. 340.

> were a course at the London Institution, for which, with the same conveniences as to laboratory and assistance, I was paid at the same rate. . . . Now twenty lectures, at the terms I have in this house, amount to £175 per annum, and therefore I should not be inclined to accept any offer under that. . . . For these reasons I wish you would originate the terms rather than I. If you could make the offer of £200 a year I would undertake them. . . . I consider the offer as a high honour, and beg you to feel assured of my sense of it. I should have been glad to have accepted or declined it independent of pecuniary motives; but my time is my only estate and that which would be occupied in the duty of the situation must be taken from what otherwise would be given to professional business.†

He was offered the job at £200 a year and held it for 20 years.

He was appointed Fullerian Professor of Chemistry at the Royal Institution in 1833, at a salary of about £100 per annum, according to the income from the endowment. His official salary as Director at the Institution remained at £100 per annum and was not changed until 1853, when it was raised to £300 per annum, but his lecture fees would have practically doubled this amount throughout that time. In 1836 he was appointed scientific adviser to Trinity House at a salary of £200 per annum and he retained this post, in which he was very interested, until 1865—that is to within 2 years of his death. In 1835 he was awarded a Civil List Pension of £300. The circumstances leading up to it were somewhat unpleasant. The pension was first proposed under the Tory government of Sir Robert Peel, but before the formalities were completed the Whigs, under Lord Melbourne, came into office. Faraday appears to have had a difficult interview in which Lord Melbourne was reported to have referred to civil pensions as a piece of gross humbug. Exactly what happened is not known for certain, but Faraday wrote to Lord Melbourne as follows:

> My Lord,
>
> The conversation with which your lordship honoured me this afternoon, including as it did, your lordship's opinion of the general character of the pensions given of late to scientific persons, induce me respectfully to decline the favour which I believe your lordship intended for me; for I feel that I could not, with satisfaction to myself, accept at your lordship's hands that which, though it has the form of approbation, is of the character which your lordship so pithily applied to it.‡

† Bence Jones, vol. II, p. 367. ‡ *Ibid.*, p. 57.

Nevertheless, the pension was granted and accepted a little later.

Faraday also received fees for consulting work though he gradually gave this up as his income from other sources grew. By about 1836 his fixed emoluments must have been some £1000 a year and his total income may well have exceeded this figure. At a time when a schoolmaster could be said to be "passing rich on £40 a year" Faraday was a wealthy man. He did not amass any fortune, however. His church preached against the accumulation of worldly riches and he spent his income as it came.

His first period of electrical researches started on 29 August 1831. It was brought to a close by a breakdown in health in 1840 which kept him away from the laboratory for 4 years. He complained of loss of memory and giddiness and had to give up his work entirely. His illness was attributed to overwork, since it affected the working of his brain, though his body at first retained its full strength. In 1841 he went with his wife, and her brother George Barnard and his wife, to Switzerland for 3 months to recuperate. He frequently did walks of 30 miles in a day, often over very rough ground, and on one occasion he records a walk of 45 miles. It has been said that the symptoms of his illness were those of mercurial poisoning and not caused by overwork at all, and from accounts of his complaint it seems that this view is not at all unlikely. He certainly worked a great deal with mercury and at the time its poisonous character was not realized. It must frequently have been spilled in his laboratory and many globules would find their way into cracks and crevices in the floor, where they would continually give off vapour. The American Standards Association lays down 1 mg of mercury vapour per 10 $m^3$ of air as the maximum allowable concentration of mercury vapour for work places. Air saturated with the vapour at 15°C would contain something of the order of 100 mg in this volume, so that unless Faraday was uncommonly fond of fresh air in his laboratory in Albemarle Street, W.1, he is quite likely to have inhaled substantially more than this maximum permissible quantity over long periods of time. Later on, his body also became affected in a way which would fit in with the hypothesis of mercurial poisoning.

Faraday returned to duty at the Royal Institution at the end of 1841 but undertook no further research except for some work on the electrification of steam issuing from an orifice, which he did in 1842. In 1844 he served on the commission investigating the Haswell Colliery disaster. He investigated conditions in the mine for himself and was amazed by the carelessness with which the miners handled the powder used for blasting. He asked his escort where they kept the powder bag. The answer was "You are sitting on it".

He resumed his experimental researches in the same year, at first continuing with the liquefaction of gases, but in 1845 he discovered diamagnetism and the rotation of the plane of polarization of light by a magnetic field. These two phenomena, and particularly the first, greatly reinforced his views about lines of force in terms of which he had accustomed himself to think. His scientific papers from this time onwards took on a much more theoretical and speculative character.

He visited Paris in the summer of 1845. He met Biot, Dumas, Arago, Babinet "and a great many others whose names and faces sadly embarrassed my poor head and memory". He continued to complain of loss of memory, confusion and giddiness. His medical advisers thought the trouble arose from "mental occupation" and "the sole remedy cessation from such occupation and head rest".

In 1846 he wrote a speculative paper under the title *Thoughts on Ray Vibrations* in which were the germs of an electromagnetic theory of light. He was not, however, able to work it out in any detail nor to calculate the velocity of propagation, a task which required the mathematical precision of Maxwell, which Faraday entirely lacked.

He took great delight in talking to children and easily won their confidence. He gave his celebrated series of juvenile lectures on "The Chemical History of a Candle" in 1849. He expressed his views on popular lectures to the Secretary of the Royal Institution as follows.

> I see no objection to evening lectures if you can find a fit man to give them. As to popular lectures (which at the same time are to be *respectable* and *sound*), none are more difficult to find. Lectures which *really teach* will never be popular; lectures which are popular will never *really teach*. They know little of the matter who think science is more easily to be taught or learned than ABC; and yet who ever learned his ABC without pain and trouble?

The year 1850 proved a busy one for Faraday's researches. The most striking piece of work, the discovery of the paramagnetism of oxygen, however, led to a blind alley—an attempt to connect terrestrial magnetism with it. In 1851 he brought a more powerful method to bear upon lines of magnetic force; he used electromagnetic induction in a wire moving across the field to measure field strengths. This method was capable of giving much more precision to his ideas of lines of force than he had achieved so far and it was a pity that it came so late in his scientific life. It convinced him of the physical reality of lines of force, and he wrote a good deal in the attempt to demonstrate this reality.

The thirtieth and last of his papers in the *Experimental Researches in Electricity* was written in 1855. In 1858 he accepted the offer of a "grace and favour" house at Hampton Court and he lived there until his death on 26 August 1867. He continued his work for Trinity House and was at sea in 1862 and 1863 when experiments were being made to compare the effectiveness of electric and oil lamps in lighthouses. He also continued with certain experimental researches but without much success. He resigned from Trinity House and also from the duties at the Royal Institution in 1865. In his last years he was seriously affected by loss of powers, particularly of his memory, but in 1865 this extended to his muscles as well. A friend from London asked him how he was. "Just waiting," he replied. He died in his chair a month short of the age of 76, and was buried in Highgate Cemetery.

CHAPTER III

# Faraday's Discovery of Electromagnetic Induction

MICHAEL FARADAY, besides making a very large number of other discoveries, many of which were of great importance to physics, contributed to the science of electrodynamics two of the basic principles upon which it was subsequently erected. The first of these was his discovery of electromagnetic induction and the laws which it follows. This, running in harness with Ampère's work on the magnetic action of electric currents, is one of the cardinal principles of the subject. It led later to the formulation of what Oliver Heaviside called the second law of circulation. Faraday's second contribution was his method of thinking in terms of lines of force and is the origin from which later field theories came to be evolved.

Faraday was, before all else, an experimentalist, and the discovery of electromagnetic induction came about as the result of a fine set of experimental researches. They comprise the first series of his *Experimental Researches in Electricity* and extracts from them form part of the example of Faraday's writings which are reproduced in the second part of this book. Exigencies of space prevent the papers being reprinted in their entirety. Throughout his *Experimental Researches* Faraday numbered all the paragraphs consecutively. They run from 1 to 3362 in the *Researches in Electricity* and the numbers will indicate what has had to be omitted.

Faraday's theoretical ideas, on the other hand, developed slowly throughout the course of his life and, though he always thought in terms of lines of force, he was continuously thinking about their nature and elaborating his conception of them. He never attained much mastery over mathematics and this, indeed, was a greater handicap to him than he probably realized. In a letter to Maxwell on 13 November 1857 he wrote:

> There is one thing I would be glad to ask you. When a mathematician engaged in investigating physical actions and results has arrived at his own conclusions, may they not be expressed in common language as fully, clearly, and definitely as in mathematical formulae? If so, would it not be a great boon to such as we to express them so—translating them out of their hieroglyphics that we might also work upon them by experiment. I think it must be so, because I have always found that you could convey to me a perfectly clear idea of your conclusions which, though they may give me no full understanding of the steps of your process, gave me the results neither above nor below the truth, and so clear in character that I can think and work from them.†

This is a plea which must have passed through the minds of most of us at one time or another, but it overlooks the contribution which mathematics makes to the process of thinking itself. Faraday's lack of mathematics effectively prevented him from ever really clarifying his ideas and it had to be left to William Thomson (afterwards Lord Kelvin) and especially Maxwell to do this for him later. Although the development of field concepts which we owe to Faraday is obviously of the greatest and most fundamental importance in the history of electricity, it is not easy to quote any of his papers which will show clearly and exactly what he was able to achieve. This part of Faraday's contribution to science will be dealt with by means of commentary in a subsequent chapter.

Throughout his life Faraday was imbued with a metaphysical belief in the "unity of the forces of nature". A similar belief in the mind of Oersted had maintained his interest in the possibility of the action of a current on a magnet over more than a dozen years of fruitless experiment. In this belief, the term "force" is, of course, used loosely and not as understood in mechanics. As early as 1824 Faraday thought that since an electric current acts upon a magnet, a magnet should react upon a current and he performed a number of unsuccessful experiments in which he expected a magnet to

† Bence Jones, vol. II, p. 387.

generate continuous currents in circuits nearby. He also tried the effect of steady currents upon neighbouring circuits, again, of course, with only negative results. It is of interest in this connection to note that the very last paper which he wrote—in 1860, towards the end of his life, but which, on the advice of Stokes was never published—was again based upon this conviction of the "unity of the forces of nature" and concerned a possible connection between electricity and gravity. It begins:

> Under the full conviction that the force of gravity is related to other forms of natural power, and is a fit subject for experiment, I endeavoured on a former occasion to discover its relation with electricity, but unsuccessfully. Under the same deep conviction, I have recently striven to procure evidence of its connection with either electricity or heat.

The experiments again gave negative results.

Faraday began his *Experimental Researches in Electricity* on 29 August 1831, and the first field which he investigated was that of electromagnetic induction. In 10 days of experimenting, spread over the period to 4 November, he completed his investigations and his paper was read to the Royal Society on 24 November, after which he went on holiday to Brighton. The first extracts which are reproduced in the second part of this book are taken from this paper. It was very clearly written and in great detail, and requires little in the way of commentary. A few of the terms employed are no longer in use. "Electricity of tension" and "common electricity" refer to what is now called static electricity. The terms "electromotor" and "voltaic arrangement" mean a battery of one or more cells. The "marked pole" of a magnet is the north-seeking pole. The terms "austral" and "boreal" poles were sometimes employed, the north-seeking pole being called "austral" because it was homologous with the south pole of the earth. This nomenclature being rather confusing, Faraday prefers the neutral term "marked pole" for that which points to the north.

The first section of the extract ends with two paragraphs on Faraday's *Electro-tonic State*. These are followed, in the *Experimental Researches*, by sixteen further paragraphs describing experiments in which a search is made for properties by which the electro-tonic state could be recognized, but the search proved fruitless. There appeared to be no properties whereby a wire in the electro-tonic state could be distinguished from another not in that state, but nevertheless, he maintained his views about it in his paper. Indeed, he wrote to his friend Phillips, from Brighton, exulting in his discovery of a new state of matter. However, by 31 December he had completed another paper—Series II in the *Experimental Researches*—in which doubts about the electro-tonic state are expressed. Paragraph 231 of that paper runs:

> The law under which the induced electric current excited in bodies moving relative to magnets, is made dependent on the intersection of the magnetic curves by the metal being thus rendered more precise and definite, seems now even to apply to the cause in the first section of the former paper; and by rendering a perfect reason for the effects produced, takes away any for supposing that peculiar condition, which I ventured to call the electro-tonic state.

However, although he thus expressed himself in writing, the idea of an electro-tonic state of matter did not completely leave his mind and he returned to the conception much later, when he was developing the idea of lines of force quantitatively. Maxwell adopted the idea and gave it a mathematical interpretation.

The electro-tonic state suggested itself to Faraday after he had discovered that the movement of magnets near a circuit, and movement or change of current in neighbouring circuits, gave rise to transient currents, and that when the movement or change was undone the same transient currents occurred in the reverse direction. He therefore formed the opinion that the induced currents were in the nature of a changing polarization of the material of which the circuit in which the currents were induced, was composed. This polarization he specifically looked upon as a condition of the particles of the material and not to electrostatic charges residing on surfaces, partly because matter in the electro-tonic state showed no attractions or repulsions and partly because in his experiments where induced currents had been recorded, he had, of course, employed complete circuits in which surface charges would not develop. His electro-tonic state thus appeared to him as a "state of tension".

In the electro-tonic state the homogeneous particles of matter appear to have assumed a regular but forced electrical arrangement in the direction of the current, which if the matter be undecomposable, produces, when relieved, a return current.†

It is interesting that as a result of his theory of an electro-tonic state of matter, Faraday was able to make two predictions, both of which have subsequently been proved to have been justified and one of which he subsequently verified himself. It is a good example of the possibility of correct predictions being made as a result of an "incorrect" theory—or at least one based upon a redundant conception. The first prediction which he made was that if a current in one circuit induces a special state in the matter of another circuit, there should be a reaction on the current in the first circuit. This he was not able to verify himself though we are now familiar with circuits coupled together in this way.

The tension of this state may therefore be comparatively very great. But whether great or small, it is hardly conceivable that it should exist without exerting a reaction upon the original inducing current and producing equilibrium of some kind. It might be anticipated that this would give rise to a retardation of the original current; but I have not been able to ascertain that this is the case. Neither have I in any other way as yet been able to distinguish effects attributable to such reaction.

The second prediction was that the material of the circuit in which the inducing current itself flows should also be thrown into the electro-tonic state.

The current of electricity which induces the electro-tonic state in a neighbouring wire, probably induces that state also in its own wire; for when by a current in one wire a collateral wire is made electro-tonic, the latter state is not rendered in any way incompatible or interfering with a current of electricity passing through it. If, therefore, the current were sent through the second wire instead of the first, it does not seem probable that its inducing action upon the second would be less, but on the contrary, more, because the distance between the agent and the matter acted upon would be very greatly diminished.

In other words, the establishment of a current should act inductively on its own circuit and Faraday thus predicted, as a result of his theory, the phenomenon of self-induction. Though the immediate experiments he carried out to test this prediction

† *Researches in Electricity*, vol. 1, p. 22.

yielded negative results, he succeeded in discovering the phenomenon experimentally 3 years later.

In his second series of *Experimental Researches in Electricity*, Faraday developed an alternative view, in which he looked upon a varying current as producing moving lines of magnetic force. This, of course, equally with the electro-tonic state, was an unobserved entity. Such moving lines of force could not be seen to move; they were simply postulated to do so. On this Faraday expressed himself as follows.

> ... the inducing wire and that under induction were arranged at a fixed distance from each other, and then an electric current sent through the former. In such cases the magnetic curves themselves must be considered as moving (if I may use the expression) across the wire under induction, from the moment at which they begin to be developed until the magnetic force of the current is at its utmost; expanding as it were from the wire outwards, and consequently being in the same relation to the fixed wire under induction as if *it* had moved in the opposite direction across them, or towards the wire carrying the current. Hence the first current induced in such cases was in the contrary direction to the principal current. On breaking the battery contact the magnetic curves (which are mere expressions for arranged magnetic forces) may be conceived as contracting upon and returning towards the failing electrical current, and therefore move in the opposite direction across the wire, and cause an opposite induced current to the first.
>
> When in experiments with ordinary magnets, the latter, in place of being moved past the wires, were actually made near them then a similar progressive development of the magnetic curves may be considered as having taken place, producing the effects which would have occurred by the motion of the wires in one direction; the destruction of the magnetic power corresponds to the motion of the wire in the opposite direction....

Faraday summarizes his final conclusion as follows:

> ... If a terminated wire move so as to cut a magnetic curve, a power is called into action which tends to urge an electric current through it; but this current cannot be brought into existence unless provision is made at the ends of the wire for its discharge and renewal.
>
> ... If a second wire move with a different velocity or in some other direction, then variations in the force exerted take place; and if connected at their extremities, an electric current passes through them.

Faraday's theory of a special electro-tonic state of matter has been described above as "incorrect". This must be taken to mean only that today we do not recognize any such special state of matter. The phenomena are interpreted entirely on the lines of

Faraday's second hypothesis, namely by the intersection of magnetic flux. Maxwell, who did not reject Faraday's electro-tonic state, pointed out that the flux of induction threading a circuit is measured by the integral of the vector potential taken round the circuit, so that if we should wish to continue to speak in terms of Faraday's electro-tonic state, then the vector potential would provide us with a measure of it. Since, however, there is nothing other than the magnetic field in which the conductor is situated with which it is possible to correlate the electro-tonic state, there would appear to be little point in trying to revive the term. The electro-tonic state means, in effect, being situated in a field of magnetic induction. This, if we wish, could be designated a special state of matter, but Faraday originally clearly meant to convey more than this by the term.

Of the electro-tonic state Maxwell wrote as follows.

> The whole history of this idea in the mind of Faraday, as shown in his published *Researches*, is well worthy of study. By a course of experiments, guided by intense application of thought, but without the aid of mathematical calculations, he was led to recognise the existence of something which we now know to be a mathematical quantity, and which may even be called the fundamental quantity in the theory of electromagnetism. But as he was led up to this conception by a purely experimental path, he ascribed to it a physical existence, and supposed it to be a peculiar condition of matter, though he was ready to abandon this theory as soon as he could explain the phenomena by any more familiar forms of thought.
>
> Other investigators were long afterwards led up to the same idea by a purely mathematical path but, so far as I know, none of them recognised, in the refined mathematical idea of the potential of two circuits, Faraday's bold hypothesis of an electro-tonic state. Those, therefore, who have approached this subject in the way pointed out by those eminent investigators who first reduced its laws to a mathematical form, have sometimes found it difficult to appreciate the scientific accuracy of the statements of the laws which Faraday, in the first two series of his Researches, has given with such wonderful completeness.
>
> The scientific value of Faraday's conception of an electro-tonic state consists in its directing the mind to lay hold of a certain quantity, on the changes of which the actual phenomena depend. Without a much greater degree of development than Faraday gave it, this conception does not easily lend itself to the explanation of the phenomena.

A great impression had been made upon the minds of physicists at the time at which Faraday began his electrical researches, by

"Arago's extraordinary experiment". This refers, of course, to the deflection of a magnet suspended above a rotating metallic disc, or the rotation of a disc mounted above a rotating magnet. Magnets of many pounds weight could be deflected in this way and the explanations so far put forward had all been in terms of magnetism induced in the rotating disc coupled with a time lag associated with the process of magnetization. Faraday saw at once that his induced currents provided a much more satisfactory explanation and also that Arago's disc provided the basis for the construction of a new electrical machine. By providing sliding contacts at the circumference and the axle of a disc spun near a magnet, Faraday constructed the first dynamo. Finally he disposed of the magnetic explanation by suspending two magnets together so that one of them was above the disc and the other below. When the two magnets were arranged so that like poles came opposite each other, Arago's effect was annulled, whereas when unlike poles were so arranged, the effect was amplified. This was the reverse of what would be expected on the theory of induced magnetism but in line with Faraday's theory of induced electric currents set up in the disc.

This little triumph over the mathematicians gave Faraday a good bit of pleasure. While he was at Brighton, after having completed his first series of experimental researches, he wrote to his friend Phillips:

> It is quite comfortable to me to find that experiment need not quail before mathematics, but is quite competent to rival it in discovery; and I am amazed to find that what the high mathematicians have announced as *the essential condition* to the rotation (in Arago's experiment)—namely that time is required—has so little foundation, that if time could by possibility be anticipated instead of being required—i.e. if the currents could be formed *before* the magnet came over the place instead of after—the effect would equally ensue.†

Such exultation was perhaps, in this case, excusable.

One of the questions to which Faraday devoted much attention —since it was in line with his idea of an electro-tonic state— concerned the possible effect which the magnetic properties pos-

† Bence Jones, vol. II, p. 10.

sessed by the material of which a circuit was composed, might have upon the currents which could be induced in it. It was only as a result of a lengthy series of experiments that he convinced himself that such properties, surprisingly, had no effect on the phenomenon at all. At first he had expected that by connecting different metals to each side of his galvanometer, connecting them also at the far end so as to complete a circuit, and then passing them through the space between the poles of a magnet together, he would obtain a differential effect. Finding himself unsuccessful in this he accordingly designed a special differential galvanometer to test this result further. The galvanometer possessed two coils to which circuits of different metals could be connected. With this instrument he succeeded in obtaining differential effects and it is with this instrument that paragraphs 205–213, which have been selected for reproduction, are concerned. Faraday's galvanometers in all these experiments are, of course, used ballistically, though at the time the difference between the significance of a ballistic throw and a steady deflection was not properly appreciated. As a result of this series of experiments he traced the differences between the metals entirely to the differences in their electrical conductivity, a result which is of fundamental importance to electrical theory, since it indicates that the electromotive force induced in a circuit is independent of the material of which the circuit is made.

As a result of his experiments Faraday finally enumerated three classes of substances according to their reaction in a magnetic field. There are first magnetic bodies like those of iron, which are acted on when stationary; then there are the electric conductors which are acted on only when moving; finally there are the indifferent bodies which are not acted on at all, either when stationary or moving. It is interesting to notice that he still seems to refer induction to a property of the substance composing the circuit. His point of view was one from which it would have appeared quite natural to postulate a special electro-tonic state to account for the phenomenon.

Faraday then went on to experiment with induced currents obtained by moving conductors in the field of the earth, which he succeeded in demonstrating both with solenoids and the rotating disc. He was led to investigate the possibility of the induction of currents in static water—in the lake at Kensington Palace—and in moving water—at Waterloo Bridge. However, he obtained no positive results with either of these investigations. At this stage his experiments on induction in the earth's field led to nothing more than the demonstration that the phenomenon occurred and added nothing substantial to his discovery of electromagnetic induction. Later on he returned to this field and made important advances, but before considering them we must turn to his next important discovery, that of self-induction. We have seen how he was actually able to predict the occurrence of the phenomenon on the basis of his theory of the electro-tonic state. His mind was thus prepared to take up the suggestion which an observation communicated to him by a Mr. William Jenkin provided. Jenkin noted the spark which occurred when an electromagnet was switched off and the shock which could be felt if the wires were held in the hands, although the battery was incapable of producing either effect directly.

Faraday could be very scathing about suggestions made by amateurs and non-professionals. Bence Jones quotes him as writing:

> The number of suggestions, hints for discovery and propositions of various kinds offered to me very freely, and with perfect goodwill and simplicity on the part of the proposers for my exclusive investigation and final honour, is remarkably great, and it is no less remarkable that but for one exception—that of Mr. Jenkin—they have all been worthless.
>
> It is quite natural that, when a man first catches sight of an analogy or relation, however imperfect it may be, he should suppose it a new and unconsidered subject, but it is very rare that the same thing has not passed before through the mind of a veteran, and been dismissed as useless. I have, I think, universally found that the man whose mind was by nature or self education fitted to make good and worthy suggestions was also the man both able and willing to work them out.
>
> The volunteers are serious embarrassments generally to the experienced philosopher.†

† Bence Jones, vol. II, p. 45.

The discovery of self-induction was recorded in the Ninth Series of the *Experimental Researches in Electricity*. The papers intervening since his first discovery of electromagnetic induction had been on identifying static and current electricity, a problem which had not been satisfactorily cleared up before, on electrolysis, into which study he introduced most of the modern terminology, and on the theory of the electric cell. The record of the discovery of self-induction is in Faraday's best style. The experiments are simple, they are clearly described and require little in the way of comment. The detailed description which Faraday gave of his experiments again makes it impossible to quote his paper in full, but it is hoped that the selection which has been made will prove sufficient to provide a reasonably connected account of the work.

Thus far the results which Faraday obtained were qualitative and at the time it would have been difficult for him to have gone much further. The laws of the electric circuit itself were by no means fully understood. It is true that the experiments of Ohm had been made in 1826 but their implications had not been fully understood nor his results entirely accepted. The electromotive force which Ohm had derived from analogy with the flow of heat had not been identified with electrostatic potential, though the identity was suspected, and further, there existed no system of generally agreed units for measurements.

By lines of magnetic force in regions near magnets or electric currents Faraday initially meant simply lines, the direction of which at any point gives the orientation of a small compass needle placed at the point. They were the lines indicated by the iron filings with which he was wont to experiment. Plates III and IV reproduce some of his drawings. However, in the end he went further than the purely qualitative results so far obtained and he was responsible for the quantitative law of induction, $E = -dN/dt$, which is named after him.

It will be observed from paragraph 3071, which is included in the extracts given here, that he also employed what he called the direction of the "magnecrystallic axis of a crystal of bismuth"

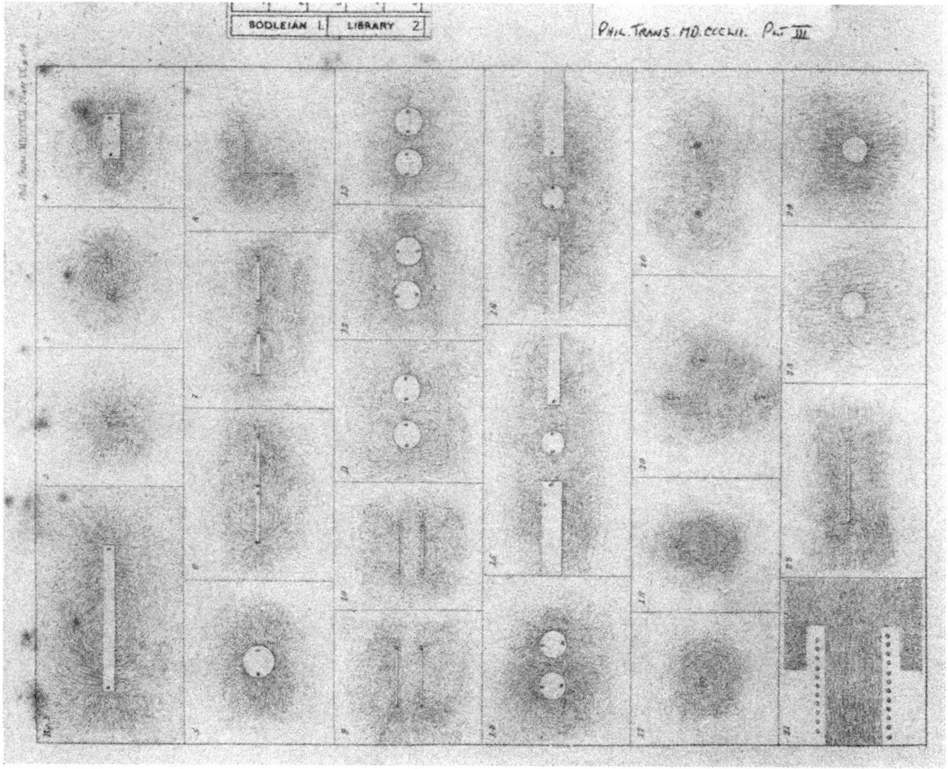


PLATE III. Faraday's Drawings of Lines of Magnetic Force as Delineated by Iron Filings. *Phil. Trans*. 1852.

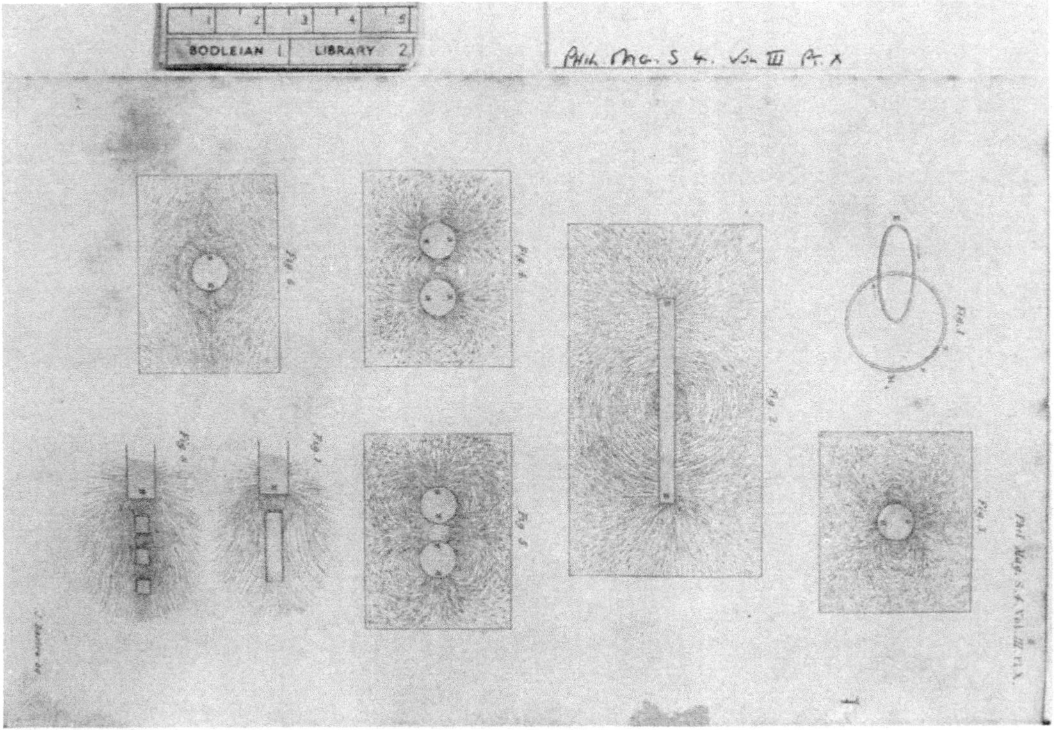


PLATE IV. Further Drawings of Lines of Force by Faraday. *Phil. Mag.*, Series 4, vol. III, 1852.

to determine the direction of the lines of force, and a word of explanation about this is, perhaps, desirable. Faraday discovered that specimens of recrystallized bismuth consisting of large crystals more or less regularly arranged, although behaving diamagnetically and being repelled from regions of high magnetic field strength, also aligned themselves in a uniform magnetic field—often with the long axis of the specimen parallel to the field. With single crystals there was an axis which always aligned itself parallel to the field. He therefore postulated a new force of nature with which matter reacted, which was neither paramagnetic nor diamagnetic in character—in that the matter was not acted upon by forces either towards or away from places of high magnetic intensity—but simply orientated in the magnetic field. To this he gave the name of "the magnecrystallic force". The effect was later investigated by Tyndall and Knoblauch, however, and shown to be a consequence of ordinary magnetic and diamagnetic action.

Some explanation of what Faraday is referring to in paragraph 3075, when he talks of the relation of the vacuum to magnetic force, is also desirable. His ideas on this question changed in the course of time. At times he appears to have looked upon a vacuum as a fourth state of matter and at other times as mere empty space. The question of the relation of a vacuum to magnetic force arose from his discovery of diamagnetism. He found that an evacuated vessel would behave as a diamagnetic substance if surrounded by a paramagnetic medium, and as a paramagnetic substance if surrounded by a diamagnetic or less paramagnetic medium. On these grounds Faraday was inclined to put the vacuum in an intermediate position in a list of substances arranged according to their para- or dia-magnetic properties. At other times he treats the vacuum as giving a zero point in his scale with paramagnetic substances on one side of it and diamagnetic substances on the other. In paragraph 3075 he appears to be thinking along the lines of the first alternative.

In making his law of induction quantitative, Faraday was faced with peculiar difficulties, for he did not work in terms of any quantitative theory of magnetism based upon a law of inverse squares of the distance, conceived as relating either to a magnetic fluid or fluids, as was currently popular at the time, or to magnetic poles or to Ampèrian molecular currents. He thus had no means of obtaining any independent quantitative measure of the strength of a magnetic field. He realized that when the lines delineated by his iron filings tended to converge on a region of space, that region would be a place where the magnetic forces were stronger, and conversely, at places from which the lines appeared to be repelled, there the forces would be weaker. This, however, he realized in a purely qualitative fashion and he had no means of rendering his ideas more precise. Indeed, he used the term force itself in a very loose fashion in his writings. It was something that dwelt along the lines of force in space. What happened in fact was that he was obliged to use the induced electromotive force in a moving conductor to specify his magnetic field quantitatively. The question arises, therefore, whether Faraday's law of induction was not, in fact, tautological. In the way in which he developed it, could it be more than a relation defining $N$ in terms of $E$?

It would not follow, of course, that if the law turned out to be tautological it would be useless in consequence. It would be possible to map out any magnetic field by means of electromagnetic induction, and when this had once been done it would then be possible to predict what electromotive force would be induced in any other circuit moving in any given manner. However, it would mean that the concept of electromotive force was logically prior to that of magnetic field strength and that any measurement of electromotive force which involved knowing the strength of a magnetic field would only be relative and not absolute.

From the beginning of his paper "On lines of Magnetic Force; their definite character and their distribution within a magnet and through space", Faraday appears to assume the possibility of representing a magnetic field quantitatively by the number of continuous lines of force drawn per unit area at right angles to the direction of the force. He assumes the continuity of his lines of force, thus, in paragraphs 3091 and 3100. It is true that he speaks in paragraph 3100 of the power of a magnet being considered to extend to any distance, "according to the recognized law", but he does not mention the fact that the continuity of his lines of force is, in fact, equivalent to the law of inverse squares, which is presumably to what the "recognized law" refers. Only a non-divergent vector can be represented quantitatively by continuous lines according to the method of Faraday, and before adopting the method it is necessary to demonstrate that the field is non-divergent. (For a discussion of this point in connection with magnetic fields generated by electric currents, see *Early Electrodynamics: The First Law of Circulation*, p. 69.)

Faraday, throughout his whole scientific life, was extremely cautious about the making of implicit assumptions and he never tired of pointing out the danger in doing so. Although it seems doubtful whether or not he realized the necessity for demonstrating the non-divergence of his magnetic field strength, since the arguments in these two paragraphs come out of their proper logical order, he did, in fact, demonstrate experimentally in paragraph 3107 that the same current was induced in a wire running from the equator to a pole of a magnet, when it was rotated about the axis of the latter through one revolution, whatever the dimensions or shape of the wire. This and the experiments which showed that no electromotive force was induced in a loop lying entirely external to the magnet and rotated about the axis of the magnet, verified that for these motions, the induced electromotive force could be interpreted quantitatively in terms of the intersection by the circuit of a system of continuous lines associated with the magnet, and lying in planes passing through the magnetic axis. Several of his experiments also indicated that the direction of such lines would be the same as those of the lines of force depicted by iron filings, and Faraday shows no hesitation in identifying the two.

That was, therefore, about as far as experiments with a magnet could take him, given the equipment at his disposal, the knowledge of the times and his general and original way of looking at things.

He repeated all these experiments a little later using a special galvanometer of low resistance, which he had constructed. This gave much greater deflections than had been obtained before, and enabled him to experiment with circuits practically all the resistance of which was then external to the galvanometer. Faraday's accounts of these experiments are not given in the extracts for want of space. They added little new of fundamental importance to what he had already discovered, but they did enable him to show more certainly that the amount of electricity "thrown into a circuit" by a given movement of a conductor in a magnetic field, varied inversely as the resistance of the circuit.

Faraday's Figs. 13 and 14.

Faraday's lines of force, which he conceived as continuous curves running through the material of a magnet, are lines of what would now be called magnetic induction and denoted by the letter *B*. His interpretation of his experiments with a magnet rotating about its axis was that the lines of force remain stationary in space while the magnet rotates within them. This, however, does not follow from his observations, since it is clear that even if the lines of force did rotate with the magnet they would not induce any resultant electromotive force into the circuits he used in his experiments, rotated in the manner he employed.

The next and final step in rendering the law of electromagnetic induction quantitative consisted in experiments carried out in a uniform field. These experiments are described in paragraphs 3177 and 3192–9. The intervening paragraphs which have been omitted merely describe the low resistance galvanometer used.

Its construction can be understood without difficulty from Figs. 13 and 14, which are reproduced. The galvanometer actually employed in these experiments had coils passing twice round each of the astatic needles and corresponded to Fig. 14. Figure 13 is of a simpler instrument used in previous experiments, and, having only one turn round each needle, its construction is more obvious from the figure.

Using modern terminology, what Faraday's experiments on induction in circuits moved in the uniform magnetic field of the earth demonstrated was that the quantity of electricity put into circulation by a given rotation varied directly as the area of the circuit rotated. Previous experiments with magnets showed that it also varied inversely as the resistance of the circuit. Putting these results together we obtain for the conclusion of his experiments that the induced electromotive force is proportional to the rate of change of flux of lines through the circuit. His results appear to be self-consistent to within about 5 per cent.

Faraday's paper on induction in the earth's magnetic field was received by the Royal Society on 31 December 1851, 20 years after the paper announcing his first discovery of the phenomenon of electromagnetic induction. It was entitled "On the Employment of the Induced Magneto-electric Current as a Test and Measure of Magnetic Forces". The title clearly indicates that he is proposing to measure the strength of a magnetic field by the method of induction, and as he makes no other experiments or proposals to attach any other quantitative meaning to the term "field strength", such a process is clearly equivalent to the definition of field strength in terms of the electromotive force induced. Two points arise for consideration. First, since he had only the constant field of the earth for his experiments, it is clear that he can only derive the law of induction in a flux cutting form. On the basis of his experiments it is impossible to say what would be the result of varying the flux through a stationary circuit. Second, the results of his experiments could be summarized by saying that when a circuit is moved in a steady magnetic field the rate of change in the number of lines of force threading it is defined from the relation

$$\frac{dN}{dt} = -E,$$

where $E$ is the electromotive force induced. This tells us nothing about the force experienced by a magnetic pole or current in the field so defined, nor anything about the strength of the field set up by a magnet or a current. A further experiment or theoretical deduction is required if the field defined as above by Faraday is to be identified with that arrived at by means of forces on magnets or currents as in the work of Ampère. We shall devote the next chapter to a discussion of these matters.

## CHAPTER IV

# The Logical Status of the Law of Electromagnetic Induction

THERE is now ample experimental evidence to support Faraday's law of the induction of currents to within an accuracy of about one in ten thousand. The law is, however, applied in circumstances where its accuracy is implied to some one in a million. How such an extrapolation is to be justified requires investigation. Further, just what experimental evidence is required to substantiate the law, also determines what use can be made of such experiments in other directions. To substantiate the law of induction of currents experimentally requires independent measurement of electromotive force and magnetic flux. There are only two basic methods for measuring an electromotive force independently of the law of induction, namely, by means of an electrostatic electrometer like the attracted disc instrument, in which the voltage can be calculated from the geometry and a measured force, and a heating experiment in which the value of the mechanical equivalent of heat is assumed. The first measurement is not capable of the necessary accuracy and the second can be taken only to about one in ten thousand. If the second is employed, two problems arise immediately. One is that which has already been mentioned, namely, how the extrapolation made in practice is to be justified; the other is, such an application rules out the use of electrical methods for the measurement of the mechanical equivalent of heat.

Faraday's law of electromagnetic induction is one of the foundation stones of both electrical theory and electrical measurement. To those who are concerned with the construction of a theory in an axiomatic deductive form for electrical science, it is obviously of importance to know the degree of validity which can be attached to the law. So, too, those who would erect the theoretical structure on the basis of measurements, on the view that these determine the meaning to be attached to the quantities and theoretical concepts employed, are equally concerned with the problem. Electrical measurements are based upon two fundamental absolute determinations. The first of these is the measurement of current by means of a current balance. It involves only the first law of circulation—i.e. Ampère's law of action of current elements. In any fundamental analysis it determines the meaning to be conveyed by the word current. The term current, in other words, becomes a portmanteau containing only the magnetic effects of the circuits said to be conveying currents; that is to say, its meaning is completely determined by the forces experienced by circuits, or parts of circuits, in the neighbourhood of other circuits, account being taken of differences of geometry by means of Ampère's formula.

The second fundamental measurement is of resistance or potential difference, which is normally carried out in standardizing laboratories by methods depending upon electromagnetic induction. A similar analysis of this operation would show, therefore, potential difference, as marked on a voltmeter or standard cell, to mean no more than a rate of change of flux and measured in units of lines per second. But the term potential difference is not defined theoretically in terms of lines per second but of work—joules per coulomb—and its general use is in terms of this theoretical definition. The connection between lines per second and joules per coulomb is provided by Faraday's law, and it is on that law that we have to rely every time we employ the standardized graduations of any voltage measuring instrument. A consideration of the logical status of the law of electromagnetic induction and of the evidence upon which we can rely in connection with it, thus enters practical as well as theoretical physics.

There are three possibilities to be considered. First, Faraday's law might be tautological—like the first two at least of Newton's laws of motion. Second, it might be possible to deduce it from the laws of the magnetic action of currents, combined with such principles of general physics as the law of conservation of energy. The third possibility is that the law is based directly upon independent experiment. If the first alternative were true it would still leave further difficulties in electrical metrology to be resolved. It would, in fact, simply define flux of induction by means of induced electromotive force and the problem of connecting flux with the action of steady currents would remain. The difficulty would be shifted without being solved. The second alternative, if it could be carried out, would resolve the difficulty completely and leave the situation neat and tidy. If, on the other hand, it is the third alternative which we have to adopt, the question then is, what are the experiments upon which we can rely with sufficient confidence to justify the use which is made of the law? In practical measurement the law is assumed accurate to better than one part in a hundred thousand. Faraday's original experiments, at best accurate to only about one part in a hundred or less, are clearly not adequate for the purpose.

We have already seen that the law of electromagnetic induction, as it left Faraday's hands, contained at least an element of tautology. He mapped his magnetic fields by induction in a moving wire. It becomes undoubtedly tautological if the experiment is simply turned round the other way and the magnetic field, thus established, is employed in the measurement of induced electromotive force. Clearly no absolute determination of potential difference is possible by induction in a magnetic field, itself determined by measurements involving potential difference. What Faraday's experiments showed was that, within the accuracy they achieved, electromagnetic induction could be attributed to the intersection by the conductor of a field, which could be represented in magnitude and direction by continuous lines—i.e. by a field that was non-divergent.

The fields used in standardizing measurements are not, of course, estimated in this way. The measurement of potential difference depends upon the use of a calculable self or mutual inductance in a Lorenz disc, bridge, or similar arrangement. (The measurements furnish resistance rather than potential difference directly, of course, but this is unimportant since potential difference is then obtained from the standard ohm and the ampere, already known from measurements with a current balance.) The standard inductance is calculated from its geometry, to which is applied Ampère's law of action of current elements. The measurement is thus equivalent to identifying Faraday's lines of force, determined by induction, with the lines of force of the magnetic field, as determined by Ampère's law. It is true that the identification is very tempting but we still need an experiment to show (a) that Faraday's lines of force follow exactly the same curves as do lines determined magnetically—few would be sufficiently hardy to maintain this identity, to the necessary degree of accuracy, as the result of scattering iron filings near a magnet or a current, suggestive as this was to Faraday and as it, indeed, remains today —and (b) that the constant of proportionality in Faraday's law has the value of unity when the strength of the field is determined from Ampère's relation instead of tautologically by electromagnetic induction.

It is often said, when questions of the validity of physical laws are raised, that they are justified not by one experiment but by a whole nexus of results deduced from them, which have been confirmed by observation. This is true but in a case such as this it can only meet part of the difficulty. This arises from the fact that the fundamental standardizing experiments are all, very naturally, based upon the most accurate measurements that it is possible to make. Indeed, present-day standardizing measurements are only possible at all when the resources of one of the main standardizing laboratories are available, and even so they are performed comparatively infrequently. The collateral evidence obtained from the confirmation of other deductions will be all of a lower order of accuracy, and so can contribute only little in comparison. Physics cannot be based upon a metaphysical belief that its laws can always be expressed in terms of small whole numbers. A rough measurement cannot be sufficient to indicate whether or not a law, for example, is one of inverse squares, and if only such measurements are available the possibility of error is clearly very great, if such results are extrapolated indefinitely. It is difficult to find any philosophical justification for the belief that nature must be essentially simple. Rather the reverse is the case. A lifetime of study devoted to a very limited field fails to exhaust its complexities. On the other hand, the laws which we use in comprehending nature are the simplest which will serve the purpose in hand. As Wittgenstein remarked in his *Tractatus*: "We make to ourselves pictures of facts." It would indeed be foolish to make these pictures more complicated than is required by the purposes for which they are to be employed, but nevertheless, pictures can be known to be accurate only within the limits to which they have been tested. Such considerations would seem to indicate that we have to look to the measurements of the standardizing laboratories, not only to furnish the values of certain basic units, but also to provide the evidence for the physical laws upon which these determinations rest. We can return to this point if necessary after having discussed the possibility of deducing Faraday's law from the known laws of the mutual action of currents and general principles of physics, since, if this proves possible, the difficulties would be resolved at one stroke and need detain us no longer.

In examining the various methods which have been suggested for the proof of Faraday's law, it is well to be on guard against accepting a demonstration of part of the law as demonstrating the whole. The law of electromagnetic induction is that the electromotive force acting round a circuit is given by

$$\frac{d}{dt}\int B.dS$$

where $dS$ is an element of a surface bounded by the circuit. This may be expressed as

$$\left(\frac{\partial B}{\partial t} + u.\text{div}\, B - \text{curl}\,(u \times B)\right) dS$$

($u$ being the velocity of an element of the circuit).

The first term represents changes in the value of the field strength; the second represents new lines of force which have begun to thread the circuit on account of the movement of the circuit with velocity $u$; the third term represents the rate of cutting of the flux by the circuit in movement.

If Faraday's flux of induction can be identified with the flux of the Ampèrian field, then it follows that div $B = 0$, and the second term in the above expression disappears. As far as the third term is concerned, Faraday did not distinguish very clearly between flux cutting and the change of flux through a circuit. He had in his mind a belief in the physical existence of the lines of force, and he imagined them to spread outwards from a wire as a current is started up in it and to collapse into it as the current stopped. This is a physical picture that requires justification since there is no *a priori* reason why flux should not be generated through a circuit without it having to come in through the boundary from outside.

To trace the story of the deduction of the law of induction to its source, we must first turn from Faraday to the continental physicists who were pursuing a different line of development. Their work, in contrast to Faraday's, was based upon the idea of action at a distance, modelled on the Newtonian theory of gravitation and employing the laws of inverse squares of the distance in electrostatics and magnetism, and Ampère's law of action of current elements in electrodynamics. The first attempt to unite the law of electromagnetic induction in a single theory with the laws of static electricity and of steady currents was that of Franz Neumann (1798–1895). It can be shown, on the basis of Ampère's law of action of current elements, that the force experienced by a circuit carrying a current $i$, can be derived from a potential function given by $i.N$, where $N$ is the number of lines of force threading the circuit, arising from currents in other neighbouring circuits or from magnets nearby (see *Early Electrodynamics, The First Law of Circulation*, p. 82). The mechanical work which has to be done in moving the circuit from a place where the number of lines of force threading it is $N_1$ to a place where the number is $N_2$, is $i(N_1 - N_2)$, the currents in all the circuits being maintained constant throughout the movement. The rate of working by the circuits at any instant is $i.dN/dt$. Neumann noticed that it is the same potential function, $i.N$, which occurs in Faraday's law of induction. Thus both the law of induction and the laws of the mechanical interaction of currents upon each other, are referable to the same potential function. This, of course, is not the same thing as deducing the law of electromagnetic induction from those of the mechanical action of currents, or vice versa. Given that currents are acted on by forces when in the neighbourhood of each other, it does not follow that they must act inductively on each other. The energy account could be balanced through the agency of the chemical actions occurring in the cells or the thermal energy involved if a thermocouple was employed, which, in the absence of the discovery of induction, would have to be incorporated in the circuits to produce the currents. This rate of chemical action would be unpredictable *a priori*.

It was in the course of his investigations leading to his potential function that Neumann was the first to introduce the vector potential, $A$. This is defined by the relation

$$\text{curl}\, A = B$$

where $B$ is the magnetic induction. Thus the flux of induction through a surface $S$ bounded by a circuit carrying a current $i$, will be

$$\int \text{curl}\, A.dS$$

which by Stokes' theorem is equal to

$$\int A.ds$$

taken round the bounding circuit $s$, and this, in turn, gives Neumann's potential function to be

$$i.\int A.ds.$$

If the magnetic induction $B$ is due to a current $i'$ in a second circuit $s'$, then

$$A = i' \int \frac{ds'}{r}$$

where $r$ is the distance from $ds'$ to $ds$ (see *Early Electrodynamics, The First Law of Circulation*, p. 66).

If this is substituted in the above expression we obtain Neumann's formula for the potential function for two circuits, which is

$$i.i' \int_s \int_{s'} \frac{ds.ds'}{r}.$$

This being symmetrical in $s$ and $s'$ shows, as we would expect, that the forces experienced by the two circuits are equal and opposite.

Neumann's work was published in 1845. At about the same time Hermann von Helmholtz (1821–94), a surgeon in the Prussian army and an accomplished musician and mathematician, who was equally prominent in both physics and physiology, attempted to deduce Faraday's law of electromagnetic induction from the mechanical forces experienced by currents, given by Ampère's law, together with the principle of the conservation of energy. This he did in a paper entitled "An Essay on the Conservation of Force" which was read before the Physical Society of Berlin on 22 July 1847. Joule's work on the mechanical equivalent of heat had been carried out around 1843 and that on the heating effect of electric currents a little earlier. Neither Joule's nor Helmholtz's efforts appear to have been received with much enthusiasm at the time by the scientific world.

Helmholtz's deduction is as follows. He considered the movement of a magnet near a circuit of resistance $R$, carrying a current $i$, maintained by a battery of electromotive force $E$. He took the potential function of the magnet in the field set up by the current $i$ to be $i.\omega$. Helmholtz entered on the expenditure side of the balance sheet for energy the rate of dissipation in Joulean heat in the resistance, $i^2R$, and the rate of working on the magnet, $i(d\omega/dt)$, while on the income side of the ledger was entered the rate of working of the battery $E.i$. Thus he arrived at the result,

$$E.i = i^2R + i\frac{d\omega}{dt}$$

giving

$$i = \frac{E - \dfrac{d\omega}{dt}}{R}.$$

If this relation is valid it will hold whatever the value of the electromotive force of the cell. Putting $E = 0$ we have,

$$i = -\frac{1}{R}\frac{d\omega}{dt}.$$

If we replace $\omega$ by the number of lines of force from the magnet which thread the circuit, as before, we obtain Faraday's law fo induction.

By putting $E = 0$ Helmholtz is virtually considering a circuit without a battery and he has thus eliminated one source of uncertainty from his calculation, namely unpredictable variations in the rate of the chemical working of the battery. There is, of course, no need to have a battery in the circuit at all, the movement of the magnet sufficing to generate the current by induction and any deduction on the lines of that of Helmholtz is the better for the elimination of that uncertain source, or sink, of energy from the calculation.

Helmholtz also attempted to calculate what happens when two electric circuits are moved relative to one another. The single relation furnished by the principle of the conservation of energy, however, is clearly insufficient to determine changes in the values of two currents and his calculations are unsatisfactory in this respect. If the explanation of magnetism on the basis of Ampèrian currents is adopted, it is possible to raise the same objection with regard to the calculation with the magnet also.

Helmholtz's calculations are also defective in another direction, as was pointed out soon afterwards by Lord Kelvin. The phenomenon of self-induction leads to a current continuing in a circuit after the electromotive force has been removed. Energy can, therefore, be obtained from the current, which cannot come directly from the battery. It must, therefore, have been stored by some mechanism in connection with the circuit. Faraday's experiment with the wire doubled back along itself, in which the effects of self-induction disappeared, indicated that the storage mechanism was not in any momentum given to and returned by ponderable electric fluid in motion in the conductors like a liquid in a pipe. The effect of the shape of the circuit on its self-inductance indicated that the seat of the storage lay in the magnetic field to which the current gave rise. Kelvin's objection to Helmholtz's calculation was that he had neglected the possibility of the storage of energy in the magnetic field. As the magnet moved, the resultant magnetic field of it and the current in the circuit changed, and though Helmholtz's calculation gave the correct result for the case of a circuit and a permanent magnet, his neglect of the possibility of the storage of energy in the field was unsatisfactory. Kelvin was able to go further and show why, nevertheless, the calculation gave the correct result. To see how this comes about and also to appreciate the point which Kelvin made, we will first endeavour to derive an expression for the energy contained in a magnetic field.

In order to obtain this expression without any consideration of electromagnetic induction, we will assume the existence of a permanent magnet. A permanent magnet, like the Newtonian particle, is a hypothetical, idealized concept. A permanent magnet does not exist but nevertheless it is a concept which can be approached sufficiently closely in practice to render it reasonable. A permanent magnet is constructed from ideally hard steel so that no change, permanent or temporary, occurs in its magnetization when it is used.

The energy contained in a magnetic field may be calculated from the work required to generate the field by the separation of two semi-infinitive permanent magnets. The calculation, which is analogous to that for estimating the energy in an electrostatic field, may be performed either in terms of magnetic poles or Ampèrian currents. We will adopt the latter alternative. According to the Ampèrian analysis the magnets may be considered as semi-infinite solenoids with current $I$ (equal to the intensity of magnetization) circulating round them, per unit length along the axis.

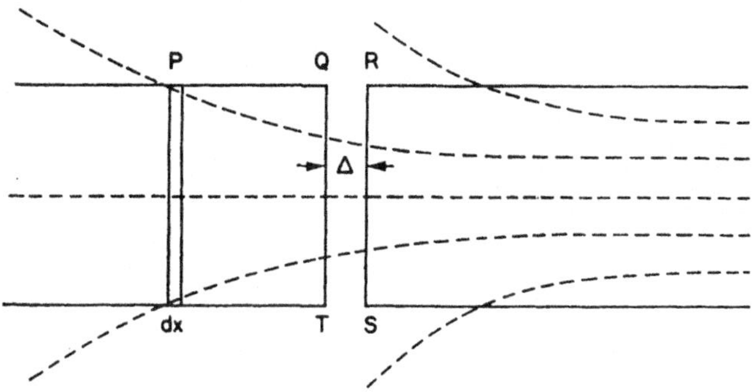


FIG. 1.

When the magnets are separated by a small distance $\Delta$, the space between them will form a flat cavity at right angles to the direction of magnetization and the field inside it will be equal to $B$, the magnetic induction. Half the lines of force of this field will arise from the right-hand magnet and will enter that on the left. Since this is made of ideally hard steel, they will emerge at various points through the sides, as would the lines of force of a solenoid situated in free space. Let $B_n$ be the component of the field normal to the side of the left-hand magnet at some point $P$ distant $x$ from the end $QT$. The component of the force parallel to the axis experienced by the slice of thickness $dx$ at $P$, arising from the interaction of this field and the Ampèrian current, will be

$$2\pi a B_n I dx$$

where $a$ is the radius of the magnet assumed cylindrical. The total force acting on the magnet $PQ$ will be

$$\int_0^\alpha 2\pi a B_n I dx = I \text{ times all the lines arising from the magnet } RS, \text{ which enter } QT$$

$$= I\frac{B}{2}A$$

where $A$ is the cross-sectional area of the magnet. The work performed in creating the field in the small volume $QRST$ is thus

$$\frac{IB}{2}$$

per unit volume. Now in the absence of any external field not arising from the magnet itself, $I$, the Ampèrian current per unit length circulating in the solenoids, is equal to $B/\mu_0 = H$, so that the energy of the magnetic field thus created in free space, is given by

$$\frac{B^2}{2\mu_0} = \frac{\mu_0 H^2}{2} = \frac{HB}{2} \text{ per unit volume.}$$

Let us assume that all magnetic fields, whether created by permanent magnets or electric currents, in free space possess this energy per unit volume, and apply the result to cases of self and mutual induction of circuits lying entirely in free space.

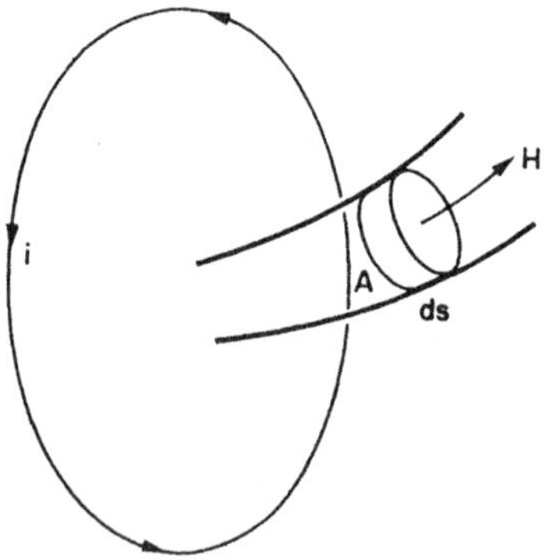


FIG. 2.

First let us take the case of an isolated circuit carrying a current $i$. If the value of the magnetic field generated at any point in space is $H$ then the total energy contained in the field will be

$$\int \frac{\mu_0 H^2}{2} d\tau$$

the integral being taken throughout space. Take as the element of volume a length $ds$ of a unit tube of induction in the field of area of cross-section $A$, so that

$$d\tau = A.ds = \frac{1}{\mu_0 H} ds.$$

Thus

$$\int \frac{\mu_0 H^2}{2} d\tau = \int \frac{H}{2} ds = \frac{i}{2}$$

per unit tube of induction threading the circuit. If there are $N$ such unit tubes the total energy in the field will be

$$\frac{Ni}{2}$$

or writing $N = Li$, where $L$ is the coefficient of self-induction as ordinarily defined, this becomes

$$\frac{Li^2}{2}$$

If the current increases, work has to be done on the system to generate the increase in field energy at the rate of

$$Li\frac{di}{dt} = i\frac{dN}{dt}.$$

The clearest case to consider is that in which the current is established in the circuit by a battery which is then suddenly removed—for example, by shorting out—and the current allowed to decay on its own. If it can be assumed that all the energy contained in the magnetic field is returned to the circuit as the current dies away, through induction in it of an electromotive force in the direction of the current, we shall have

$$Ei = -i\frac{dN}{dt}$$

or

$$E = -\frac{dN}{dt}.$$

This is probably the most direct and the simplest deduction which can be provided of Faraday's law though not a general deduction. There is no battery in the circuit to complicate matters by constituting an unknown source of energy, nor is there magnetic material in the neighbourhood, as in Helmholtz's calculation with the magnet. We are, therefore, in a good position to examine the assumptions which are included in the deduction which incidentally gives the law in the flux changing form and not that of flux cutting, the latter being, on the whole, easier to deal with theoretically. In examining the assumptions made in the demonstration, we need not concern ourselves with the principle of the conservation of energy, which is one of the most widely used principles of physics, and which, if not true, would involve much more than the induction of currents. This is not the place for a discussion of the validity of this law. Probably the first assumption which appears to need consideration will be that of the possibility of the permanent magnet assumed in calculating the energy in a magnetic field. This is a concept which shares the limitations possessed by all the theoretical concepts employed in physics. It possesses postulated properties, and experiment alone can decide whether such a concept can be realized in practice. No theoretical deduction can ever eliminate the need for experiment. However, the postulate of the permanent magnet does not appear an unreasonable one to make. Its use amounts to the decision to ascribe other energy changes found in the case of actual magnets to causes other than the energy of the magnetic field—to hysteresis for example. Reliance on permanent magnetism in laying the foundations of electrical science, however, is a little unfortunate from another point of view. Subsequent development has been in the direction of making current electricity basic and interpreting magnetism in terms of electric currents. As Ampère pointed out, this is the only way in which it is possible to unify the sciences of magnetism and electricity. The assumption of ideally permanent magnets is equivalent to assuming that the Ampèrian currents are unaffected by external magnetic fields, and against this are the phenomena of diamagnetism, which appear to demand the opposite assumption for their explanation.

Then there is, of course, the assumption that the energy employed in separating the two semi-infinite permanent magnets resides in the field which is generated in space between them. This is the basis of the whole of field theory. The work done in separating the magnets is performed on the magnets themselves, and when it is regained as the magnets are allowed to come together again, it appears as work done by the magnets. Even when circuits are moved about in the field of a permanent magnet, none of the energy of the field is ever transferred to them, as we shall see shortly. If it were so, then Helmholtz would not have obtained the correct result as he did. The continental physicists, working with theories of action at a distance, did not have this concept of energy contained in the field, available to them.

Another assumption is that as the field collapses in the case of the current, all the energy of the field is transferred to the circuit from which it was derived—in the absence of other circuits in the neighbourhood. If other circuits exist in the neighbourhood, however, it is impossible to say, *a priori*, how the energy will distribute itself among them, as we shall now see. It is also assumed that the circuit and its field form a simple conservative system. It is impossible to forecast whether or not some of the energy will be radiated away either as the currents are established or as they decay. If it is, the law of induction could be different to what the simple conservative view would indicate.

Let us now pass to the case of two circuits in which currents $i_1$ and $i_2$ flow respectively, again situated in free space.

Suppose the field generated by the current $i_1$ in the first circuit

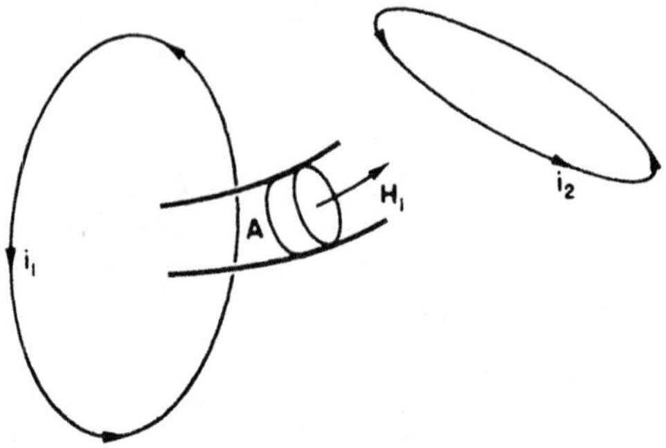


FIG. 3.

to be represented by the vector $\boldsymbol{H}_1$ ($H_{1x}$, $H_{1y}$, $H_{1z}$) and that generated by the current $i_2$ in the second, by $\boldsymbol{H}_2$ ($H_{2x}$, $H_{2y}$, $H_{2z}$). If the resultant field is represented by $\boldsymbol{H}$ we shall have

$$\boldsymbol{H}^2 = (\boldsymbol{H}_1 + \boldsymbol{H}_2)^2$$
$$= \boldsymbol{H}_1^2 + \boldsymbol{H}_2^2 + 2\boldsymbol{H}_1 . \boldsymbol{H}_2.$$

Integrating throughout space, as before, we have

$$\int \mu_0 \frac{\boldsymbol{H}_1^2}{2} d\tau = \frac{1}{2} L_1 i_1^2$$

and

$$\int \mu_0 \frac{\boldsymbol{H}_2^2}{2} d\tau = \frac{1}{2} L_2 i_2^2$$

where $L_1$ and $L_2$ are the coefficients of self-induction of the two circuits.

To integrate $$\int \boldsymbol{H}_1 . 2 \boldsymbol{H} d\tau$$

take as the element of volume an element of a unit tube of induction in the field of $H_1$. We shall have

$$\int \mu_0 (H_{1x}H_{2x} + H_{1y}H_{2y} + H_{1z}H_{2z})\, d\tau$$
$$= \int (H_{1x}H_{2x} + H_{1y}H_{2y} + H_{1z}H_{2z}) \frac{ds}{H_1}$$
$$= \int (l_1 H_{2x} + m_1 H_{2y} + n_1 H_{2z})\, ds$$

where $l_1$, $m_1$, $n_1$ are the direction cosines of $\boldsymbol{H}_1$. The expression in the brackets is therefore the resolved part of $\boldsymbol{H}_2$ in the direction of $\boldsymbol{H}_1$, that is of $d\boldsymbol{s}$. The integral is, therefore,

$$\int \boldsymbol{H}_2 d\boldsymbol{s}$$

taken round all the unit tubes of induction in the field $\boldsymbol{H}_1$. Now $\boldsymbol{H}_2$ is the magnetic field generated by the second circuit in the absence of the first, so that the integral vanishes for all tubes of $\boldsymbol{H}_1$ which do not thread the second circuit, and for those that do it equals $i_2$, the current in it. Thus the integral becomes

$$i_2 N$$

where $N$ is the number of unit tubes of induction arising from the first circuit which thread the second. Writing this $Mi_1$, in the usual way, where $M$ is the coefficient of mutual induction as ordinarily defined, the integral becomes

$$Mi_1 i_2.$$

Thus the total energy in the field will be

$$\tfrac{1}{2}L_1 i_1^2 + Mi_1 i_2 + \tfrac{1}{2}L_2 i_2^2.$$

To attempt to discuss electromagnetic induction as before, let the batteries in both circuits be shorted out and the currents allowed to decay. Let $E_1$ be the electromotive force arising in the first circuit by induction from variations in both the currents $i_1$ and $i_2$, and similarly let $E_2$ be that in the second, arising in the same way. Then if the circuits are rigid and fixed so that the coefficients $L_1$, $L_2$, and $M$ remain constant, the principle of the conservation of energy provides the relation:

$$i_1 \left( E_1 + L_1 \frac{di_1}{dt} + M \frac{di_2}{dt} \right) + i_2 \left( E_2 + L_2 \frac{di_2}{dt} + M \frac{di_1}{dt} \right) = 0.$$

Faraday's law is that

$$E_1 + L_1 \frac{di_1}{dt} + M \frac{di_2}{dt}$$

and

$$E_2 + L_2 \frac{di_2}{dt} + M \frac{di_1}{dt}$$

are separately zero. This, however, cannot be made to follow from the single relation, which is all that the law of the conservation of energy is able to provide. This is a serious limitation. It means that a deduction based solely on the principle of the conservation of energy cannot be made for the general case where electric currents and conducting material are spread throughout space. It is impossible to say, on the basis of this single principle, how the energy in the field will be distributed among the various conducting circuits when the currents are no longer maintained.

It is, however, possible to consider certain cases where it is obvious what the currents in the circuits must be, relative to each other, by considerations of symmetry. Let us take the case of two equal circuits at a great distance from each other, so that their mutual action is negligible. Let us suppose that each possesses a

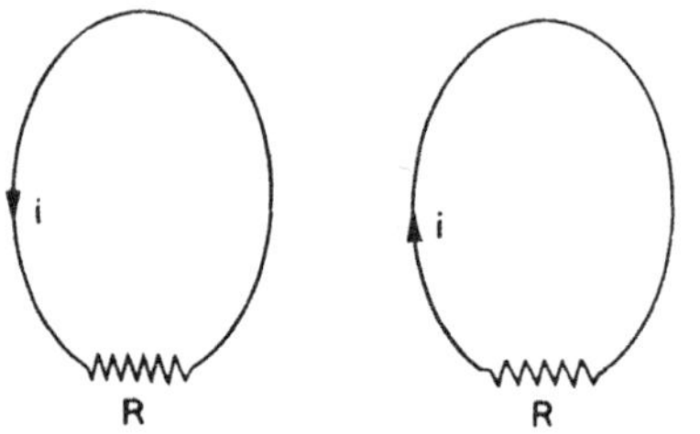


FIG. 4.

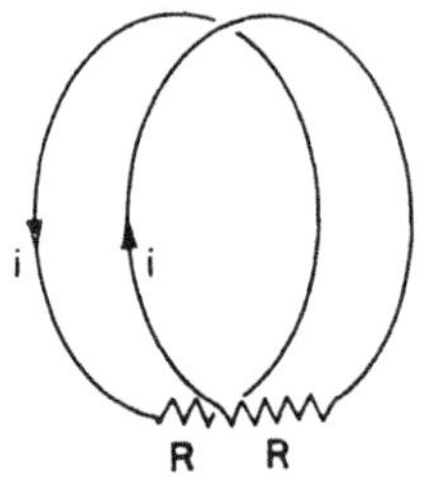


FIG. 5.

resistance $R$, and let equal and opposite currents be started in each, for example, by the sudden removal of a magnet. If the current flowing in either circuit is $i$, each circuit will possess energy $\frac{1}{2}L_i^2$, stored in its magnetic field, which will be dissipated in the resistance if nothing is done to maintain the currents and thus the fields. Let us now suppose that the circuits are made to approach each other. Each will induce in the other a current which will be in the same direction as that already existing. We may, therefore, suppose that the circuits are made to approach each other at such a rate as to maintain the currents constant. The currents being in opposing directions in the two circuits a force of repulsion will be experienced by each, and work will have to be performed to make the circuits approach. When the circuits coincide a total amount of work amounting to $W = Mi^2$ will have had to be done, where $M$ is the coefficient of mutual induction between the circuits in the final position, as follows from the Ampèrian potential function. When the circuits coincide $M = L$ and the fields of the two currents will annul each other. Let $E$ be the electromotive force developed in either circuit during the movement. Finally, all the energy contained in the original fields together with the equivalent of the work done must have been dissipated in the two resistances. Thus,

$$2Ei = W + Li^2$$

$$= 2W.$$

Had the circuits been allowed to repel each other instead of being made to approach, the currents would, of course, have quickly decayed. But suppose the circuits had started close together and the currents been maintained constant by means of batteries, while the circuits repelled each other to a distance; then these batteries would have had to supply energy equal to twice the external work performed by the circuits during the repulsion, the difference between what the batteries supplied and the external work being stored in the field.

This is a perfectly general result. Let two circuits, possessing a coefficient of mutual induction $M$, have currents $i_1$ and $i_2$ maintained in them by batteries, while they are allowed to move relatively to each other and perform external work. The external work performed while $M$ changes from $M_1$ to $M_2$ will be $(M_2 - M_1)\, i_1 i_2$. The increase in the energy stored in the field, at the same time, will also be $(M_2 - M_1)\, i_1 i_2$. The batteries have, therefore, to supply twice this amount of energy to maintain the currents constant, a result that was first given by Lord Kelvin.

It might, at first sight, be supposed that when a permanent magnet is placed in the neighbourhood of an electric current, there would be a term in the expression for the energy of the resultant field depending upon the overlapping of the two components, as was the case with the two circuits, since the energy depends on the square of the field strength. Kelvin also showed that this could not be so. To see how this comes about let us consider again the case of the two circuits with equal and opposite currents flowing in them and made to approach each other, but now let us replace one of the circuits by a permanent magnet possessing an equivalent distribution of magnetism to the current which it replaces. If the energy contained in the field of the permanent magnet alone is $\lambda$, the initial energy, when the two bodies are widely separated, will be $\frac{1}{2}Li^2 + \lambda$. When the magnet has been brought into close proximity with the current, the same amount of work will have been performed as before and the magnetic field will, again, have been destroyed. We might, therefore, expect to be able to switch off the current without any of the energy from the self-induction of the circuit being manifest. When the current is switched off, however, the magnetic field of the permanent magnet will have to be regenerated, since the current which annulled it will no longer be there. The regeneration of the field of the magnet, as the current is switched off, will induce into the circuit an electromotive force, which will be the same as though the current which the magnet replaced had been switched on, and thus will tend to maintain the current which is being switched off. Since the field which is being regenerated is equal to that of the current, energy equal to $\frac{1}{2}Li^2$ will appear in the circuit, just as though the current had been switched off in the absence of the magnet. The balance sheet for the energy is, therefore, as follows:

| | |
|---|---|
| Initial energy of the fields | $= \frac{1}{2}Li^2 + \lambda$ |
| Work performed | $= W$ |
| | $W + \frac{1}{2}Li^2 + \lambda$ |

| | |
|---|---|
| Final energy of field | $= \lambda$ |
| Energy dissipated on switching off | $= \frac{1}{2}Li^2$ |
| Balance available for dissipation in resistance during approach | $= W$ |
| | $W + \frac{1}{2}Li^2 + \lambda$ |

All the external work, therefore, which was performed by the agent bringing the magnet and the circuit together, must have been dissipated in the resistance. The result is precisely the same as though the energies in the two fields remained the same, in spite of their overlap, and the external work performed was all used in the generation of the induced e.m.f.

The result is obviously general. If, as the result of the movement of a permanent magnet near a circuit, against the electrodynamic forces, $N$ lines from the magnet thread the circuit, then energy equal to $\frac{1}{2}Li^2 - Ni$ will be returned from the remaining field on switching off, together with $Ni$ when the original field of the magnet is restored. The energy balance indicates that all the external work will appear as heat dissipated in the resistance of the circuit. Kelvin was thus able to explain the result, which is surprising at first sight, that Helmholtz, who took no account of the energy that must be stored in the magnetic field, yet obtained the correct result when he considered the movement of a permanent magnet near a circuit.

The result is no longer true when induced magnetism is taken into account. External work is performed by a current as a piece of magnetically susceptible material approaches it and also the energy in the field is increased. Experiments of the kind that Faraday performed are sufficient to show that the effects of self-induction are considerably increased by the presence of soft-iron cores, and that soft iron is more effective in this way than hard steel. In the absence of permanent magnetism a battery would be necessary to maintain the current and the battery would supply energy equivalent to the external work performed together with the extra energy stored in the field.

There is scope for the exercise of a certain amount of ingenuity in devising deductions of Faraday's law on the lines of Helmholtz's calculation, which are free from the necessity to assume the complete absence of induced magnetism, and from Kelvin's objection concerning the neglect of the storage of energy in the fields. Both these objections can be eliminated if the device is such that the fields generated remain stationary. A simple arrangement of this kind would be Faraday's disc dynamo. In this a copper disc is rotated between the poles of a permanent magnet and sliding contacts made with the circumference and the axle. If the radial current is $i$, the radius of the disc $a$, and the rate of rotation $n$ revolutions per second, the rate of working will be

$$\pi a^2 iBn$$

where $B$ is the magnetic induction produced by the magnet. If an e.m.f. $E$ is induced in the circuit

$$Ei = \pi a^2 iBn$$

and

$$E = \pi a^2 Bn$$

which is the rate of cutting of the lines of force by a radius of the disc. While in this arrangement Faraday's idea of flux cutting is fairly apparent, both circuit and magnet remain in the same positions relative to each other and the interpretation to be given to the rate of change of flux threading the circuit becomes obscure. All that is happening to the circuit is that one piece of metal is continually being replaced by another. The experiment gives no clue to what e.m.f. would be induced by a changing flux threading a stationary circuit, arising, say, from variations in current in a second circuit at rest relative to the first. It is true that Faraday imagined the lines of force to be moving in such cases, but this is a picture useful to explain the results of experiments *post hoc*, but not a fact which could be asserted *a priori*. Also we are not entirely freed, in this demonstration, from the necessity to assume permanent magnetism, since if the field were generated by a current in a second circuit (as in the Lorenz disc, for example) the battery required to drive this current could be the repository of the energy used in driving the disc.

A different kind of theory was provided by Weber in 1846. He assumed that an electric current consisted of a stream of positive electricity moving with velocity $u$ in one direction, and an equal stream of negative electricity moving with the same velocity in the opposite direction. He then investigated the law of force between charges (depending upon their velocities and accelerations as well as upon their distance apart) which must exist if Ampère's law of action of current elements is to be true. From this he obtained their mutual energy and, by an argument similar to Neumann's, the law of induction. In outline his theory is as follows.

Let one current consist of quantities of positive and negative electricity, $\lambda$, moving with equal and opposite velocities $u$, and let the same quantities for the second current be $\lambda'$ and $u'$, so that the first current $i = 2\lambda u$ and the second $i' = 2\lambda' u'$. Using electromagnetic units in place of the so-called electrodynamic ones employed by Ampère, his law of action of current elements is that two elements $ids$ and $i'ds'$ repel each other with a force

$$F = ii'\,ds\,ds'\left(\frac{2}{r}\frac{d^2r}{ds\,ds'} - \frac{1}{r^2}\frac{dr}{ds}\frac{dr}{ds'}\right).$$

Consider the action of the positive electricity in the first current, first on the positive and then on the negative electricity of the second current. We have for the two positive electricities,

$$\left(\frac{dr}{dt}\right)_1 = u\frac{dr}{ds} + u'\frac{dr}{ds'}$$

and

$$\left(\frac{d^2r}{dt^2}\right)_1 = u^2\frac{d^2r}{ds^2} + 2uu'\frac{d^2r}{ds\,ds'} + u'^2\frac{d^2r}{ds'^2}.$$

For the positive electricity in the first current and the negative electricity in the second, the same expressions will hold if the sign of $u'$ is changed. Thus,

$$\left(\frac{dr}{dt}\right)_2 = u\frac{dr}{ds} - u'\frac{dr}{ds'}$$

and

$$\left(\frac{d^2r}{dt^2}\right)_2 = u^2\frac{d^2r}{ds^2} - 2uu'\frac{d^2r}{ds\,ds'} + u'^2\frac{d^2r}{ds'^2}.$$

Combining these expressions we find for half the force between the current elements,

$$\frac{F}{2} = \frac{\lambda\lambda'\,ds\,ds'}{r^2}\left[r\left\{\left(\frac{d^2r}{dt^2}\right)_1 - \left(\frac{d^2r}{dt^2}\right)_2\right\} - \frac{1}{2}\left\{\left(\frac{dr}{dt}\right)_1^2 - \left(\frac{dr}{dt}\right)_2^2\right\}\right]$$

The action of the negative electricity in the first current on the positive and negative electricities in the second produces a similar expression, so that we see that the force $F$ is the same as would arise if two charges of electricity $e$ and $e'$ repelled each other with a force

$$\frac{ee'}{r^2}\left\{r\frac{d^2r}{dt^2} - \frac{1}{2}\left(\frac{dr}{dt}\right)^2\right\}.$$

To this must be added the force of electrostatic repulsion

$$\frac{ee'c^2}{r^2},$$

so that the complete expression for the force between two charges is, according to Weber's theory,

$$\frac{ee'c^2}{r^2}\left\{1 + \frac{r}{c^2}\frac{d^2r}{dt^2} - \frac{1}{2c^2}\left(\frac{dr}{dt}\right)^2\right\}.$$

This force is the same as though the particles possessed a mutual energy

$$E = \frac{ee'c^2}{r}\left\{1 - \frac{1}{2c^2}\left(\frac{dr}{dt}\right)^2\right\}.$$

[For we have the rate of diminution of the electrical energy corresponding to a change in $r$ is

$$-\frac{dE}{dt}=\frac{ee'c^2}{r^2}\frac{dr}{dt}-\frac{ee'}{2r^2}\left(\frac{dr}{dt}\right)^3+\frac{ee'}{r}\frac{d^2r}{dt^2}\frac{dr}{dt}$$

$$=\frac{ee'c^2}{r^2}\left\{1+\frac{r}{c^2}\frac{d^2r}{dt^2}-\frac{1}{2c^2}\left(\frac{dr}{dt}\right)^2\right\}\frac{dr}{dt}.$$

This, by the principle of virtual work, must equal the rate of working of the electrical forces, which is $F(dr/dt)$, which agrees with the above expression for the force.]

If $\bar{v}$ and $\bar{v}'$ denote the velocities of the charges,

$$E=\frac{ee'c^2}{r}\left(1-\frac{(\bar{r}.\bar{v}-\bar{r}.\bar{v}')^2}{2c^2r^2}\right).$$

Summing for the interaction of charges of both signs in each current this becomes

$$E=\frac{4ee'}{r^3}(\bar{r}.\bar{v})(\bar{r}.\bar{v}')=ii'\frac{(\bar{r}.ds)(\bar{r}.ds')}{r^3}$$

since

$$i=2ev.$$

Writing the vector potential

$$A=\int\frac{(\bar{r}.ds')\,\bar{r}}{r^3}$$

the integral being taken round the circuit $s'$, the total energy thus becomes

$$E=\int iAds$$

where the integral is taken round the circuit $s$. This integral equals the flux of induction threading the circuit $s$, multiplied by the current flowing in it. Faraday's law for mutual induction follows from this expression when $i'$, the circuit $s'$ or the circuit $s$ round which the integral is taken, is varied.

Weber's theory, however, breaks down completely if an electric current is not composed of equal amounts of positive and negative electricity possessing equal and opposite velocities. Though, at first sight, this might appear to be a minor conceptual detail, in fact it is not possible to remove this difficulty without destroying the theory. The theory is obviously much too detailed and fragile to provide a satisfactory basis for the laws of electromagnetic induction. It was, however, the first of the theories in which the phenomena of electrodynamics are explained in terms of forces between charges depending upon their separation and velocities. It was felt at the time that the basis of such a fundamental law as that of electromagnetic induction should not be made to rest upon so detailed a conception of the constitution of an electric current. In spite of the progress which has been made with electron theories of conduction, the same remains true today. The basis of the theory should not need to be changed with every change in the view of the nature of the electric current, and the concept of electron flow in conductors that is necessitated by the Hall effect, for example, is far removed from the simple ideas which alone can be available for a deduction of Faraday's law.

We give Maxwell's treatment of the problem on pages 219 *et seq.* His methods are much more general and in keeping with an analysis of phenomena that have been made the foundation of electromagnetic theory, than deductions based on detailed assumptions about the constitution of the electric current. Maxwell's theory is worked out in terms of the generalized mechanics of Lagrange, in which it is possible to eliminate the need for detailed knowledge of the nature of the electric medium. Faraday took special care, at least for most of his time, to avoid assumptions about what constitutes an electric current, but though sceptical of the views held at his time, as Maxwell pointed out, he was not always successful in this aim. In electrolysis in particular, a subject which engaged a good deal of Faraday's attention, it proved very difficult to avoid altogether the idea of transport of charge. However, he insisted that of the nature of the electric current he remained ignorant. He defined what he meant by the word current in paragraph 283 of the *Experimental Researches*.

> By *current*, I mean anything progressive, whether it be a fluid of electricity or two fluids moving in opposite directions, or merely vibra-

tions, or, speaking still more generally, progressive forces. By *arrangement* I understand a local adjustment of particles, or fluids or forces not progressive. Many other reasons might be urged in support of the view of a *current* rather than an *arrangement* but I am anxious to avoid stating unnecessarily what will occur to others at the moment.

Later in paragraph 516 he wrote:

> I have sought amongst the various experiments quoted in support of these views, or connected with electrochemical compositions or electric currents, for any which might be considered as sustaining the theory of two electricities rather than that of one, but have not been able to perceive a single fact which could be brought forward for such a purpose; or admitting the hypothesis of two electricities, much less have I been able to perceive the slightest grounds for believing that one electricity in a current can be more powerful than the other, or that it can be present without the other, or that one can be varied or in the slightest degree affected, without a compensating variation in the other. . . .

Later still, in paragraph 1617, he said "The word current is so expressive in common language that when applied in the consideration of electrical phenomena we can hardly divest it sufficiently of its meaning or prevent the mind from being prejudiced by it". The reminder is, perhaps, not altogether irrelevant in these more sophisticated days when we have become familiar with talking about the actual nature of an electric current, without the evidence for our views being always present to the mind. The recondite properties of many common conductors should be a warning against the erection of our theory upon over-simplified views of the nature of the electric current.

Maxwell's treatment requires little in the way of comment. He points out the assumptions he has to make in the course of his presentation of his theory. They are probably as little detailed as it is possible to make them if the experimental results are to be deduced. He requires additional experimental evidence concerning the ordinary inertia of electricity, and the experiments which he makes to provide it he admits are fairly rough. He finds it necessary to assume that energy of the nature of kinetic energy can be stored in space and, he says, this involves something moving somewhere, whose motion depends upon the strengths of the currents, and since its value must depend upon the products of each pair of currents, the moving matter cannot be confined to the conductors carrying the current. Electromotive forces act on electricity and generate kinetic energy, just as mechanical forces generate mechanical kinetic energy, but this kinetic energy is stored outside the conductors in which the electromotive forces act. He has also to assume the continuity of electricity.

It will be noticed that he refers to the hypothesis that an electric current consists of equal flows of positive and negative fluids, as being due to Fechner. This is correct. Weber, who, as we have already noticed, made the same assumption, took it from Fechner. Maxwell, however, prefers not to make such a detailed assumption, even though it could explain the lack of inertial effects of electricity which he was at pains to investigate experimentally.

Maxwell's theory, like the others, links the phenomena of electromagnetic induction with those of the magnetic action of currents by means of certain assumptions. Neumann linked the two sets together by means of his potential function. Weber linked them by means of his hypothesis of equal and opposite flows of positive and negative electricity. Maxwell's principal hypothesis is that of kinetic energy in space.

The passage from Maxwell's *Treatise*, which we reproduce herewith, ends with the quantities he represents by $L$, $M$, etc., as unspecified coefficients. His expression for the mechanical action between two currents, however, serves to identify them. The laws of the mechanical action of currents, established as a result of Ampère's investigations, show that the mechanical force on a circuit carrying a current $i_1$, arising from a second current $i_2$ in another circuit, is given by

$$\frac{d}{dx}\int B_2 dS$$

where $B_2$ is the magnetic induction produced by the second circuit and the integral is taken over any surface bounded by the first circuit. This induction is defined from Ampère's law to be

$$i_2 \text{ grad } \phi_2$$

where $\phi_2$ is the solid angle subtended by the second circuit. Thus Maxwell's $M$ is identical with the mutual inductance of the two circuits defined in the usual way. On this view the definition of magnetic induction is removed from experiments with magnets and, since such experiments play no part in modern measurement, this is as it should be. The properties of magnetic materials, however, will naturally continue to figure in electrical theory in an important role, but they can now be placed in a subsidiary position. They can be considered either on views built up on the basis of magnetic poles or, alternatively, of Ampèrian currents.

There remains the view of electromagnetic induction taken in the theory of relativity. This is that, as physical phenomena must be invariant with regard to uniform motion, the distinction between an electric and a moving magnetic field disappears. Electromagnetic induction, being a first-order effect, the complications arising from the details of relativity theory may be omitted from a first consideration. Suppose a conductor to be transported across a field of induction $\bar{B}$ with velocity $\bar{v}$. Let us suppose that the conductor contains electricity that is free to move; it may be of one sign or the other or both. The moving positive electricity will be equivalent to a current and it will experience a force per unit charge equal to the vector product of $\bar{B}$ and $\bar{v}$. The moving negative electricity will experience a force in the opposite direction. Either process is equivalent to the introduction of an electric field of strength $(\bar{B} \times \bar{v})$, which is Faraday's flux-cutting rule. The usual difficulties encountered when the flux through a fixed circuit is changing remain, as with the other deductions of the flux-cutting rule, and we have to assume again that electricity within conductors moves in a simple way in response to electric and magnetic fields.

We have now surveyed most of the theories which have been developed in the attempt to link the law of electromagnetic induction with the forces exerted by steady currents upon each other. All depend upon the acceptance of certain assumptions. Most of these would have been looked upon as plausible, *a priori*, though later knowledge would make it appear to have been extremely hazardous to predict the consequences of electricity inside conductors being subjected simultaneously to electric and magnetic fields. The movement of electricity within conductors has proved to be far from simple, the collective behaviour of electrons, for example, sometimes involving an effective negative mass, so that motion is in the opposite direction to that expected. Flux cutting, i.e. the movement of a conductor in a magnetic field, seems to be rather more amenable to theoretical discussion than does the case of a fixed circuit in a varying magnetic field. Faraday looked upon the two as identical. This involves another assumption which, however, with his belief in the physical existence of lines of force, Faraday almost takes for granted. The non-divergence of magnetic induction was demonstrated by him only for stationary fields, and what happens to the lines in a varying field cannot be foretold from considerations of fixed fields alone. It might or might not be legitimate to look upon lines of force as spreading out from a wire or collapsing into it as the current varies and thus cutting through neighbouring conductors. For this and other reasons, it may appear to some preferable to regard Faraday's law as independent and based directly upon experiment. If so, it is necessary to return to our first question and ask what are the experiments upon which we can rely. What are required for this purpose are experiments accurate to better than one part in a hundred thousand. Before embarking on this enquiry it is well to become clear about a difference which exists between establishing the first law of circulation (i.e. Ampère's law of action of current elements) and the second (Faraday's law of induction of currents). To verify Ampère's law it is necessary to measure the forces between circuits of different geometries and this is, in effect, done by the standardizing laboratories, by means of the various electrodynamometers employed for determining the ampere. Although the different current balances employed were not designed with this end in view, sufficient variety exists (e.g. between the Rayleigh type of balance most frequently employed and the Pellat electrodynamometer used by R. L. Driscoll†) to furnish a verification of the law to a high degree of accuracy. These experiments, in fact, test the geometrical factor in Ampère's law. There is also, of course, a constant of proportionality to be determined but in the case of current this can be agreed upon by definition. Such a process, however, can only occur once in the theory of current electricity. As soon as the ampere has been fixed, the volt is determined in terms of work. While it might be comparatively easy to show to a high degree of accuracy, that Faraday's law is

$$E = -K\frac{dN}{dt}$$

(which is as far as his own rough experiments went) to determine the constant, $K$, it is necessary to employ an independent measure of $E$.

The first candidate for examination would be Lorenz's experiment for the standardization of the ohm. A copper disc is rotated in the field of a current $I$ circulating in a set of coils, and the resulting induced electromotive force, picked up by sliding contacts touching the edge and the axle, is balanced against that developed by the current $I$ in the resistance $R$ being standardized. If the disc makes $n$ revolutions per second and $M$ is the coefficient of mutual induction between the coils and the area traced out by the sliding contacts,

$$RI = MIn$$

or

$$R = Mn.$$

The resistance is thus determined in terms of a coefficient of mutual inductance and a rate of rotation. The rate at which the disc cuts the flux of induction running through it will vary as the flux is varied, so that if the experiments showed that the value obtained for the resistance was independent of the strength of the current $I$, it would constitute a verification of Faraday's law in the flux cutting form, to the extent that the induced electromotive force would be shown to be proportional to the rate of cutting

† *Journal of National Bureau of Standards*, **60**, 287 (1958).

of flux. It would not, however, determine the constant of proportionality.

The Lorenz method has now been largely replaced by bridge measurements with a self or mutual inductance, the value of which can be calculated from its geometry and Ampère's law of action of current elements. The calculation involves only the first law of circulation. Agreement between these bridge measurements and those made with the rotating disc would provide evidence of the required degree of accuracy supporting Faraday's law in the flux changing form, if the law in the form of flux cutting could be assumed established. If we rely on Lorenz's disc for this, the constant of proportionality remains undetermined. If we rely on theory we fall back on the assumption of permanent magnetism.

A third line on the problem would be provided by determinations of the ohm in terms of a calculable capacitance, which have been carried out in Australia and America. Agreement between the standards thus produced with those obtained from the induction experiments would furnish accurate confirmation of the law of induction. These measurements, however, entail a knowledge of the ratio of the electromagnetic to the electrostatic units, and if this were to be obtained from the measurements of the velocity of light, it would amount to an assumption of the Maxwellian theory, the basis of which we are attempting to establish. The evidence in favour of Faraday's law, which is an important component in that theory, would, therefore, at best be indirect and at the worst, circular.

A fourth measurement, which has been carried out with an accuracy approaching that which is required, is provided by heating experiments. These are normally looked upon as furnishing a value for the mechanical equivalent of heat, but the possibility of using them the other way round, to determine a resistance in absolute measure, has been explored by the German standardizing laboratory. Agreement of the standards thus produced with those obtained from induction experiments would again furnish evidence to support Faraday's law. It seems likely that the accuracy obtainable by this comparison would be about an order of magnitude less than the best results obtainable by the other methods.

There is one important point about the use of heating experiments which does not apply to the other verifications. The heating experiments are the only ones which can relate the induced electromotive force absolutely to units defined in terms of work. Agreement among the values of the ohm obtained by Lorenz's disc run at various speeds or by standard inductances operated at different frequencies, can only establish Faraday's law in a relative manner, that is to say, they would demonstrate that the induced electromotive force was proportional to the rate of change of flux. They could not furnish the constant of proportionality.

At this stage it is only possible to leave the matter as an open question. All the experiments just listed were carried out with the object of establishing a standard of resistance. They were not performed with any idea of confirming the basic laws of electrodynamics, so that factors such as variations in speed of rotation of the Lorenz disc, and the frequency of the current employed in the bridge method, were decided upon in the light of other considerations, and were not attended to with this end in view. Both theoretical and practical approaches thus leave a good deal to be desired.

CHAPTER V

# Faraday as the Founder of Field Theory

IT IS given to few men in the realm of physics to develop original views which stand apart from the accepted pattern. Most must be content to extend a pattern already existing, but Faraday's ideas were peculiarly his own. For him the term force did not have its customary meaning. Much of his work was inspired by a conviction concerning "the unity of the forces of nature". He did not originate this idea. It had inspired many others, such as Oersted in their investigations. The idea was vague; it did not mean that the various phenomena of nature could be traced to a common origin but rather that they were interconnected. It corresponded to the desire to develop theories which are more and more general that has lain behind much research in theoretical physics, and continues to do so.

The concept upon which Faraday centred so much of his thinking in electricity was that of lines of force. On this Tyndall wrote:

> Let it be remembered that Faraday entertained notions regarding matter and force altogether distinct from the view ordinarily held by scientific men. Force seemed to him an entity dwelling along the line in which it is exerted. The lines on which gravity acts between the sun and the earth seemed figured in his mind as so many elastic strings: indeed, he accepts the assumed instantaneity of gravity as the expression of the enormous elasticity of the "lines of weight". Such views, fruitful in the case of magnetism, barren as yet in the case of gravity, explain his efforts to transform this latter force. When he goes into the open air and permits his helices to fall, to his mind's eye they are tearing through the lines of gravitating power, and hence his hope and conviction that an effect would and ought to be produced. It must ever be borne in mind that Faraday's difficulty in dealing with these conceptions was at bottom the same as that of Newton; that he is, in fact, trying to overleap this difficulty, and with it probably the limits prescribed to the intellect itself.

The idea of lines of magnetic force was suggested to Faraday by the linear arrangement of iron filings when scattered over a magnet. He speaks of, and illustrates by sketches, the deflection, both convergent and divergent, of the lines of force, when they pass respectively through magnetic and diamagnetic bodies. These notions of concentration and divergence are also based on the direct observation of his iron filings. So long did he brood upon these lines; so habitually did he associate them with his experiments on induced currents, that the association became "indissoluble" and he could not think without them. "I have been so accustomed", he writes, "to employ them and especially in my last researches, that I may have unwittingly become prejudiced in their favour, and ceased to be a clear sighted judge. Still I have always endeavoured to make experiment the test and controller of theory and opinion; but neither by that nor by close cross-examination in principle, have I been made aware of any error involved in their use".†

From an early time Faraday's thoughts turned to the relation of matter and space. In his commonplace book there is the following entry in 1818. "Bodies do not act where they are not. Query: is not the reverse of this true? Do not all bodies act where they are not and do any of them act where they are?" This is a question to which he frequently returned. His final solution seemed to him to lie in a theory of matter on the lines of Boscovitch's centres of force. In his notes he wrote:

Final brooding impression that particles are only centres of force; that the force or forces constitute matter; that therefore there is no space between the particles distinct from the particles of matter; that they touch each other just as much in gases as in liquids or solids; and that they are materially penetrable, probably even to their very centres.

In a letter to Richard Taylor in 1844 he wrote:

The view now stated of the constitution of matter would seem to involve necessarily the conclusion that matter fills all space, or, at least, all space to which gravitation extends (including the sun and its system); for gravitation is a property of matter dependent upon a certain force, and it is this force which constitutes the matter. In that view matter is not merely mutually penetrable, but each atom extends, so to say, throughout the whole of the solar system, yet always retaining its own centre of force. This, at first sight, seems to fall in very harmoniously with Mossotti's mathematical investigations and reference of the phenomena of electricity, cohesion, gravitation, etc., to one force in matter; and also again with the old adage "matter cannot act where it is not".

† Bence Jones, p. 274.

Faraday's concept of lines of force originated in magnetism and he was at first inclined to picture them as being propagated only through the agency of matter. He had no very clear ideas about what happened in a vacuum but as he seemed to think that no vacuum had, in fact, ever been made, what happened in one appeared to him to be unknown. Indeed, as the above quotation shows, he seemed to doubt whether a vacuum could exist, though on this his opinion varied from time to time. As far as magnetic matter was concerned he said:

> It requires mere observation to be satisfied that when a magnet is acting upon a piece of soft iron, the iron itself, by the condition which its particles assume carries on the force to distant points, giving it direction and concentration in a manner most striking.

Although he wrote this in 1845 at a time when his thinking had become more completely crystallized, he had already, much earlier, transferred the idea to the case of electrostatics. Although the lines of force could not be observed in electric fields with the ease that was possible in magnetic fields, he traced them in imagination. Indeed he does describe one experiment with small pieces of silk to delineate their paths. All matter could be regarded as capable of electrification even though only certain bodies were magnetic, and he was thus able to transfer the observation concerning the magnet and the piece of soft iron to the electrostatic case and build, by analogy, a structure for the electric field. Faraday held the idea to the end that electric fields were only propagated through the medium of matter and that induction was transferred from point to point by an electric polarization of the dielectric—a name he gave to the substance, including a vacuum, through which an electric field was propagated. One of the first topics to which he found these ideas could be usefully applied, was that of electrolysis. He developed the theory in Series XI of his *Experimental Researches* in which he wrote in November 1837:

> When I discovered the general fact that electrolytes refused to yield their elements to a current when in the solid state, though they gave them forth freely if in the liquid condition, I thought I saw an opening to the

elucidation of inductive action, and the possible subjugation of many dissimilar phenomena to one law. For let the electrolyte be water, a plate of ice being coated with platina foil on its two surfaces, and these coatings connected with any continued source of the two electrical powers, the ice will charge like a Leyden arrangement, presenting a case of common induction, but no current will pass. If the ice be liquefied, the induction will fall to a certain degree, because a current can now pass; but its passing is dependent upon a *peculiar molecular arrangement* of the particles, consistent with the transfer of the elements of the electrolyte in opposite directions, the degree of discharge and the quantity of elements evolved being exactly proportioned to each other. Whether the charging of the metallic coating be effected by a powerful electrical machine, a strong and large voltaic battery or a single pair of plates, makes no difference in principle, but only in degree of action. Common induction takes place in each case if the electrolyte be solid, or if fluid, chemical action and decomposition ensue, providing opposing actions do not interfere; and it is of high importance occasionally thus to compare effects in their extreme degree, for the purpose of enabling us to comprehend the nature of an action in its weak state, which may be only sufficiently evident to us in its strongest condition. As therefore, in the electrolytic action, *induction* appeared to be the *first* step, and *decomposition* the *second* (the power of separating these steps from each other by giving the solid or fluid condition to the electrolyte being in our hands); as the induction was the same in its nature as that through air, glass, wax, etc., produced by any of the ordinary means; and as the whole effect in the electrolyte appeared to be an action of particles thrown into a peculiar or polarised state, I was led to suspect that common induction itself was in all cases an *action of contiguous particles* and that electrical action at a distance (i.e. ordinary inductive action) never occurred except through the influence of the intervening matter.

The respect which I entertain towards the names of Epinus, Cavendish, Poisson, and other most eminent men, all of whose theories I believe consider induction as an action at a distance and in straight lines, long indisposed me to the view I have just stated; and though I always watched for opportunities to prove the opposite opinion, and made such experiments occasionally as seemed to bear directly on the point, as, for instance, the examination of electrolytes, solid and fluid, whilst under the influence of polarised light, it is only of late and by degrees, that the extreme generality of the subject has urged me still further to extend my experiments and publish my views. At present I believe ordinary induction in all cases to be an action of contiguous particles consisting in a species of polarity instead of being an action of either particles or masses at sensible distances; and if this be true, the distinction and establishment of such a truth must be of the greatest consequence to our further progress in the investigation of the nature of electric forces.

Faraday's theory of electrolysis was, in fact, that the particles of the electrolyte were first "aligned or polarized" in the electric field. Then neighbouring particles divided and changed partners thus destroying the polarization. This was followed by a reorientation and so on. One element of the compound was thus carried in one direction and the second in the other.

Another fact which impressed itself on Faraday's mind was that, in his view, an isolated positive or negative charge could not exist on its own.

> In association with this [he wrote] the impossibility under any circumstances, as yet, of absolutely charging matter of any kind with one or the other electricity only, dwelt in my mind, and made me wish to search for a clearer view than any I was acquainted with, of the way in which electrical powers and the particles of matter are related; especially in inductive action, upon which almost all others appeared to rest.†

This statement may strike one as rather odd since nothing appears easier than to charge a body with one kind of electricity. What was in Faraday's mind was the fact that positive and negative electricities always appeared together in equal amounts. A charge on one body was always accompanied by an equal and opposite charge elsewhere.

> It was in attempts to prove the existence of electricity separate from matter, by giving an independent charge of either positive or negative power only, to some one substance, and the utter failure of all such attempts, whatever substance was used or whatever means of exciting or *evolving* electricity were employed, that first drove me to look upon induction as an action of the particles of matter, each having *both* forces developed in it in exactly equal amount.‡

Electric charges were like magnetic poles; they appeared on the surfaces of bodies because of the polarization of the matter between them. In the case of magnetism the polarized matter was the material of the magnet. In the case of electric charges it was the dielectric between opposite charges. The electric charge was thus little more than the termination of the lines of electric force. "Though I shall use the terms positive and negative by them I merely mean the termini of such lines."§ The view offered a ready

† *Experimental Researches*, 1163.
‡ *Experimental Researches*, 1168.
§ *Diary*, 3 Aug. 1837.

explanation of why electric charge resides only on the surface of conductors. Faraday built a cube with sides 12 ft long, covered in a network of wire and insulated.

> I went into the cube and lived in it [he said] using lighted candles, electrometers and all other tests of electrical states. I could not find the least influence upon them, or indication of anything particular given by them, though all the time the outside of the cube was powerfully charged, and large sparks and brushes were darting off from every part of its outer surface.†

On the other hand, charging bodies inside the cube produced charges on the walls of the chamber.

Throughout his long discussion of the question of lines of force, Faraday attached immense importance to the fact that they could be curved and not straight. He seemed to see in this an argument supporting his theory of polarization of a medium, as opposed to a theory of action at a distance. He looked upon gravity as the paradigm of action at a distance and for some reason imagined that the lines of force of a gravitational field were necessarily straight. There is, it is true, so far as is known, nothing corresponding to induction of charge in the case of gravitation and the presence of one mass does not affect the attraction arising from another, but the lines of force of the resultant field of a number of gravitating masses are certainly not straight, and it is difficult to see any reason underlying this argument by Faraday.

The idea that electric fields were propagated through the polarization of ordinary matter was queried by Dr. Hare, then Professor of Chemistry at the University of Pennsylvania. He wrote to Faraday about it as follows.

> . . . As soon as I commenced the perusal of your researches on this subject, it occurred to me that the passage of electricity through a vacuum, or highly rarified medium, as demonstrated by various experiments, and especially those of Davy, was inconsistent with the idea that ponderable matter could be a necessary agent in the process of electrical induction. I therefore inferred that your efforts would be primarily directed to a re-examination of that question.
>
> If induction, in acting through a vacuum, be propagated in right lines, may not the curvilinear direction which it possesses, when passing

† *Experimental Researches*, 1173.

through "dielectrics", be ascribed to the modifying influences which they exert?

If as you concede, electrified particles on opposite sides of a vacuum can act upon each other [Faraday had said that by "contiguous" is meant "those that are next"] wherefore is the received theory of the mode in which the excited surface of a Leyden jar, induces in the opposite surface a contrary state, objectionable? ....

It appears to me that there has been an undue disposition to burden the matter, usually regarded as such, with more duties than it can perform. Although it is only with the properties of matter that we have direct acquaintance, and the existence of matter rests upon a theoretical inference that since we perceive properties, there must be material particles to which the properties belong; yet there is no conviction which the mass of mankind entertains with more firmness than that of the existence of matter in that ponderable form, in which it is instinctively recognised by people of common sense. Not perceiving that this conviction can only be supported as a theoretic deduction from our perception of properties, there is a reluctance to admit the existence of other matter, which has not in its favour the same instinctive conception although theoretically similar reasoning would apply. But if one kind of matter be admitted to exist because we perceive properties, the existence of which cannot otherwise be explained, are we not warranted if we notice more properties than can reasonably be assigned to one kind of matter, to assume the existence of another kind of matter?

Independently of the considerations which have led some philosophers to suppose that we are surrounded by an ocean of electric matter, which by its redundancy or deficiency is capable of producing the phenomena of mechanical electricity, it has appeared to me inconceivable that the phenomena of galvanism and electro-magnetism, latterly brought into view, can be satisfactorily explained without supposing the agency of an intervening imponderable medium by whose subserviency the inductive influence of currents or magnets is propagated. ...

Having devised so many ingenious experiments tending to show that the received ideas of electrical induction are inadequate to explain the phenomena without supposing a modifying influence in intervening ponderable matter, should there prove to be cases in which the results cannot be satisfactorily explained by ascribing them to ponderable particles, I hope that you may be induced to review the whole ground, in order to determine whether the part to be assigned to contiguous ponderable particles, be not secondary to that performed by the imponderable principles by which they are surrounded.

Hare is not here criticizing Faraday's field approach but merely wishes to avoid making it depend upon the properties of ordinary matter. Faraday's reply on this point is not very satisfactory. He repeats what he has said about the meaning he proposes to attach to the word "contiguous". "I mean by contiguous particles those which are next to each other, not that there is no space between them." Comparing "ordinary induction" with

> . . . a very hypothetical case, that of a vacuum, I have said nothing in my theory [he says] which forbids that a charged particle in the centre of a vacuum should not act on the particle next to it, though that should be half an inch off. With the meaning which I have carefully attached to the word contiguous I see no contradiction here in the terms used, nor any natural impossibility or improbability in such an action. Nevertheless *ordinary* induction is to me an action of contiguous particles, being particles at insensible distances: induction across a vacuum is not an ordinary instance, and yet I do not perceive that it cannot come under the same principle of action.

This appears to be a quibble about the meaning of words and in this argument Faraday seems to make the worst of both worlds. At this point he is not prepared to challenge publicly the view of a vacuum as mere empty space. His theory of "ordinary induction" is that of action by particles at "insensible distances". Induction across half an inch of vacuum can be nothing else than action at a distance and in any case it is not clear how the conception of action at insensible distances would remove any difficulties that there may be in the idea of action at a distance. Action across a vacuum reinstates them completely. Nevertheless, the tenacity with which Faraday clung to his idea that the lines of force were properties of matter rather than a disembodied aether, may have enabled him to take the initial step and to think about them more freely. He might otherwise have felt reluctant to assign the properties to them which he did and to put forward suggestions of their "physical" existence. For example, he looked upon his lines as being in a state of tension longitudinally and compression laterally; they had a tendency to shorten and to thicken.

> The direct inductive force, which may be conceived to be exerted in lines between the two limiting and charged conducting surfaces, is accompanied by a lateral or transverse force equivalent to a dilatation or repulsion of these representative lines; or the attractive force which exists amongst the particles of the dielectric in the direction of the induction is accompanied by a repulsion or diverging force in the transverse direction.†

† *Experimental Researches*, 1297.

Faraday developed a theory of the electric current very similar in general pattern to his theory of electrolysis. As a preliminary he presents a model of an insulating dielectric.

> If the space round a charged globe were filled with a mixture of an insulating dielectric, as oil of turpentine or air, and small globular conductors as shot, the latter being at a little distance from each other so as to be insulated, then these would, in their condition and action resemble what I consider to be the condition and action of the particles of the insulating dielectric itself. If the globe were charged, these little conductors would all be polar; if the globe were discharged they would all return to their normal state, to be polarised again upon the recharging of the globe.

The theory of the electric current

> assumes that all the *particles*, whether of insulating or conducting matter, are as wholes, conductors. That not being polar in their normal state, they can become so by the influence of neighbouring charged particles, the polar state being developed at the instant, exactly as in an insulated conducting *mass* consisting of many particles. That the particles when polarised are in a forced state, and tend to return to their normal natural condition. That being as wholes conductors, they can readily be charged, either *bodily* or *polarly*. That particles which being contiguous are also in the line of inductive action can communicate or transfer their polar forces to one another, *more or less* readily. That those doing so less readily require the polar force to be raised to a higher degree before this transference or communication takes place. That the *ready* communication of forces between contiguous particles constitutes conduction, and difficult communication insulation.†

Faraday was faced with greater difficulty in picturing a process by which magnetic fields can be established throughout space. He wished here again to employ his mechanism of the handing on of properties by contiguous particles, but the particles of ordinary matter did not appear to be magnetic. In his early experiments he had been unable to detect any influence on a magnetic field causing electromagnetic induction, by the interposing of plates o most substances, such as shellac, sulphur or copper. ". . . it may be asked what is the relation of the particles of insulating bodies, such as air, sulphur or lac, when they intervene in the line of magnetic action?" Here Faraday falls back upon the electro-tonic state of his first researches.

† *Experimental Researches*, 1669–75.

> The answer to this is at present merely conjectural. I have long thought there must be a particular condition of such bodies corresponding to the state which causes currents in metals and other conductors; and considering that the bodies are insulators one would expect that state to be one of tension. I have by rotating non-conducting bodies near magnetic poles and poles near them, and also by causing powerful electric currents to be suddenly formed and to cease around and about insulators in various directions, endeavoured to make some such state sensible, but have not succeeded. Nevertheless, as any such state must be of exceedingly low intensity, because of the feeble intensity of the currents which are used to induce it, it may well be that the state may exist, and may be discoverable by some more expert experimentalist, though I have not been able to make it sensible.
>
> It appears to me possible, therefore, and even probable, that magnetic action may be communicated to a distance by the action of the intervening particles, in a manner having a relation to the way in which the inductive forces of static electricity are transferred to a distance; the intervening particles assuming for the time more or less of a peculiar condition, which (though with a very imperfect idea) I have several times expressed by the term *electro-tonic.* I hope I will not be understood that I hold the settled opinion that such is the case. I would rather in fact have proved the contrary, namely, that magnetic forces are quite independent of the matter intervening between the inductric and inducteous bodies, but I cannot get over the difficulty presented by such substances as copper, silver, lead, gold, carbon and even aqueous solutions, which though they are known to assume a peculiar state whilst intervening between the bodies acting and acted upon, no more interfere with the final result than those which have as yet had no peculiarity of condition discovered in them.†

If, with Maxwell, we identify Faraday's electro-tonic state with the vector potential of the magnetic field, the latter's theory is equivalent to endowing this potential with the function of a cause. The vector potential would then be the origin of the field, rather than a mere measure of it. It is a function of the currents generating the field and, if causes are to be sought, the idea is not altogether unreasonable. The vector potential would also have to be endowed with self-reproductive properties so that it would spread from point to point in space. Again this presents no greater difficulty than the spread of a magnetic field from its generating currents. However, though the concept of an electro-tonic state could perhaps be of some value as providing a mechanism which could be pictured, from the present point of view, both electro-

† *Experimental Researches*, 1728–9.

tonic state and vector potential are redundant. The field could equally well be looked upon as being caused by the currents directly. Though he got very close to it, Faraday's ability to create mental pictures never really took him far enough to give him a good theory of field propagation.

In December 1845 Faraday announced his discovery of diamagnetism. He found that all bodies, or nearly all, show an effect of one kind or another when placed in a magnetic field. These results contributed substantially to his general picture of magnetic phenomena and deserve mention on that account. Of particular interest in this connection are his experiments with magnetic solutions. He made up solutions of "proto-sulphate of iron" of various strengths and found that the behaviour of bodies in the magnetic field depended upon the medium in which they were placed. If a diamagnetic body were placed in a solution more strongly diamagnetic than itself, it behaved as a paramagnetic body, experiencing forces which directed it to the stronger parts of the field. This fact directed his attention to the conditions obtaining in a vacuum.

At first he looked upon diamagnetism as a phenomenon separate from paramagnetism—he drew no distinction between paramagnetism and ferromagnetism.

> All the phenomena resolve themselves into this, that a particle of such matter, when under magnetic action, tends to move from stronger to weaker places or points of force. When the substance is surrounded by lines of magnetic force of equal power on all sides, it does not tend to move, and is then in marked contradistinction with a linear current of electricity under the same circumstances.
>
> This condition and effect is new, not *only* as it respects the exertion of power by a magnet over bodies previously supposed to be indifferent to its influence, but it is *new* as a magnetic action, presenting us with a second mode in which the magnetic power can exert its influence. These two modes are in the same general anti-thetical relation to each other as positive and negative in electricity: and the diamagnetic phenomena are the more important, because they extend largely, and in a new direction, that character of duality which the magnetic force already, in a certain degree, was known to possess.
>
> All matter appears to be subject to the magnetic force as universally as it is to the gravitating, the electric and the chemical or cohesive forces...†

† *Experimental Researches*, 2418–20.

Faraday's first theory of diamagnetism was a polar one as follows:

> theoretically, an explanation of the movement of diamagnetic bodies, and all the dynamic phenomena consequent upon the actions of magnets on them, might be offered in the supposition that magnetic induction caused in them a contrary state to that which it produced in magnetic matter; i.e. that if a particle of each kind of matter were placed in the magnetic field both would become magnetic, and each have its axis parallel to the resultant magnetic force passing through it; but the particle of magnetic matter would have its north and south poles opposite, or facing towards the contrary poles of the inducing magnet, whereas with the diamagnetic particles the reverse would be the case; and hence would result approximation in the one substance, recession in the other.
>
> Upon Ampère's theory, this view would be equivalent to the supposition, that as currents are induced in iron and magnets parallel to those existing in the inducing magnet or battery wire; so in bismuth, heavy glass and diamagnetic bodies, the currents induced are in the contrary direction. This would make the currents in the diamagnetics the same in direction as those which are induced in diamagnetic conductors at the *commencement* of the inducing current; and those in magnetic bodies the same as those produced at the *cessation* of the same inducing current. No difficulty would occur as respects non-conducting magnetic and diamagnetic substances because the hypothetical currents are supposed to exist not in the mass, but round the particles of matter.†

(By the term diamagnetic conductor Faraday here means merely a conductor lying in the magnetic field. This was Faraday's original meaning of the term diamagnetic. A dielectric was a substance through which an electric field passed and a diamagnetic was one through which a magnetic field passed. The term diamagnetic originally had no reference to movement in a magnetic field and it was employed later in this way because most diamagnetics set at right angles to the field. He later introduced the term paramagnetic and then diamagnetic became its antithesis.)

This polar theory had no strong appeal for Faraday and even in the paper in which he described it he set out tentative considerations leading in a different direction. At this stage he had not fully succeeded with experiments with air and gases but nevertheless he wrote:

> The effects with air are, however, in these respects precisely the same as those which were obtained with the solutions of iron of various

† *Experimental Researches*, 2429–30.

strengths when *all* bodies belonged to the magnetic class, and when the effect was evidently due to the greater or smaller degree of magnetic power possessed by the solutions. ... It is easy to perceive that if all bodies were magnetic in different degrees, forming one great series from end to end, with air in the middle of the series, the effects would take place as they do actually occur. Any body from the middle part of the series would point equatorically in bodies above it and axially in those beneath it. ...

Such a view would also make mere space magnetic, and precisely to the same degree as air and gases. Now though it may very well be that space, air and gases, have the same general relation to magnetic force, it seems to me a great additional assumption to suppose that they are all absolutely magnetic, and in the midst of a series of bodies, rather than to suppose that they are in a normal or zero state. For the present, therefore, I incline to the former view, and consequently to the opinion that diamagnetics have a specific action antithetically distinct from ordinary magnetic action, and have thus presented us with a magnetic property new to our knowledge.†

The experiments with gases made so far had been unsuccessful because any magnetic property they possessed had been masked by those of the containing vessels. However, in August 1850 Faraday described a new differential torsion balance which enabled this difficulty to be overcome. The new apparatus led immediately to the discovery of the paramagnetism of oxygen. This opened up once again the question of the nature of the action of diamagnetic substances and once again the properties of free space led to difficulties of conception.

Before determining the place of zero amongst magnetic and diamagnetic bodies, we have [said Faraday] to consider the true character and relation of space free from any material substance. Though we cannot procure a space perfectly free from matter, one can make a close approximation to it in a carefully prepared Torricellian vacuum. Perhaps it is hardly necessary for me to state that I find both iron and bismuth in such vacua perfectly obedient to the magnet. From such experiments, and also from general observations and knowledge, it seems manifest that the lines of magnetic force can traverse pure space, just as gravitating force does, and as electrical forces do; and therefore space has a magnetic relation of its own, and one that we shall probably find hereafter to be of the utmost importance in natural phenomena. But this character of space is not of the same kind as that which, in relation to matter, we endeavour to express by the terms magnetic and diamagnetic. To confuse these together would be to confound space with matter and to trouble all the conceptions by which we endeavour to understand and work at a

† *Experimental Researches*, 2438 and 2440.

progressively clearer view of the mode of action and the laws of natural forces. It would be as if, in gravitation or electric forces, one were to confound the particles acting on each other with the space across which they are acting, and would I think, shut the door to advancement. Mere space cannot act as matter acts, even though the utmost latitude be allowed for the hypothesis of an ether; and admitting that hypothesis, it would be a large additional assumption to suppose that the lines of magnetic force are vibrations carried on in it; whilst as yet, we have no proof or indication that time is required for their propagation, or in what respect they may in general character assimilate to or differ from, the respective lines of gravitating, luminiferous or electric forces.

Neither can space be supposed to have those circular currents round points diffused through it, which Ampère's theory assumes to exist around the particles of ordinary magnetic matter, and which I had for the moment supposed might exist in the contrary direction round the particles of diamagnetic matter. . . .†

Two months later the next series of *Experimental Researches* appeared, dated October 1850, and in this was developed a field description of magnetic action.

The remarkable results given in a former series of these Researches respecting the powerful tendency of certain gaseous substances to proceed either to or from the central line of magnetic force, according to their relation to other substances present at the same time . . . have, upon consideration, led me to the idea, that if bodies possess different degrees of *conducting power* for magnetism, that difference may account for all the phenomena; and further that if such an idea be considered, it may assist in developing the nature of magnetic force. I shall therefore venture to think and speak freely on this matter for a while, for the purpose of drawing others into a consideration of the subject; though I run the risk in doing so, of falling into error through imperfect experiments and reasoning. As yet, however, I only state the case hypothetically, and use the phrase *conducting power* as a general expression of the capability which bodies may possess of effecting the transmission of magnetic force; implying nothing as to how the process of conduction is carried on. . . .

If a medium having a certain conducting power occupy the magnetic field, and then a portion of another medium or substance be placed in the field having a greater conducting power, the latter will tend to draw up towards the place of greatest force, displacing the former. Such at least is the case with bodies that are freely magnetic, as iron, nickel, cobalt and their combinations, and such a result is in analogy with the phenomena produced by electric induction. If a portion of still higher conducting power be brought into play, it will approach the axial line and displace that which had just gone there; so that a body having a certain amount of conducting power, will appear as if attracted in a

† *Experimental Researches*, 2787–8.

medium of weaker power, and as if repelled in a medium of stronger power by this differential kind of action. . . .

I have already said, in reference to the transference onwards of magnetic force, that pure space or a vacuum, permits that transference, independent of any function that can be considered as of the same nature as the conducting power of matter; and in a manner more analogous to that in which the lines of gravitating force, or of static electric force, pass across mere space. Thus as respects those bodies which, like oxygen, facilitate the transmission of this power more or less, they class together as magnetic or paramagnetic substances; and those bodies, which, like olefiant gas or phosphorus, give more or less obstruction may be arranged together as the diamagnetic class. . . .

If such be a correct general view of the nature and differences of paramagnetic and diamagnetic substances, then the internal processes by which they perform their functions can hardly be the same, though they might be similar. Thus they *may* have circular electric currents in opposite directions, but their distinction can scarcely be supposed to depend upon the difference of force of currents in the *same* direction.†

Faraday then considers the effect of the displacement of his lines of force, into paramagnetic and out of diamagnetic bodies, on the strengths of the field in the neighbouring areas. The tendency of diamagnetic bodies to go to the weaker parts of the field is shown to lead two such bodies, placed side by side, to repel each other, the field strength between them being intensified, and that of paramagnetic bodies to go to the strongest parts, likewise leads to a mutual repulsion since the field strength between them will be diminished. A diamagnetic and a paramagnetic body, on the other hand, would be expected to attract each other. The movement in all three cases can be deduced from Faraday's figure on which Fig. 6 is based.

Dated a year later—October 1851—appeared a further series in the *Experimental Researches*, "On Lines of Magnetic Force; their definite character; and their distribution within a magnet and through space". In this the use of lines of force in the description of the phenomena of magnetism is explained and the paper contains a comparison of this view with that of action at a distance. Faraday says:

How the magnetic force is transferred through bodies or through space we know not: whether the result is merely action at a distance, as in the

† *Experimental Researches*, 2797–8, 2802–3.

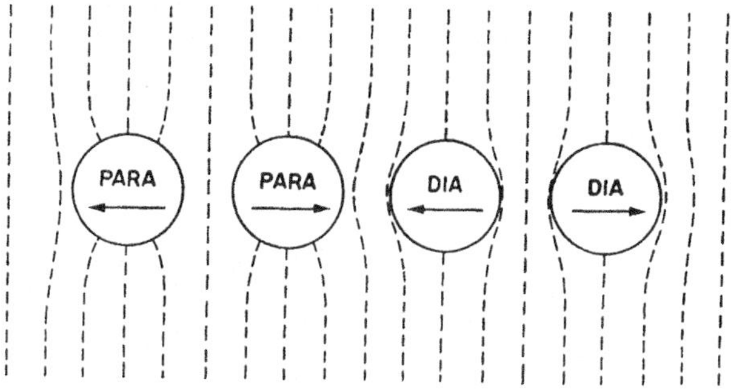


FIG. 6.

case of gravity; or by some intermediate agency, as in the case of light, heat, the electric current, and (as I believe) static electric action. The idea of magnetic fluids, as applied by some, or of magnetic centres of action, does not include that of the latter kind of transmission, but the idea of lines of force does. Nevertheless, because a particular method of representing the force does not include such a mode of transmission, the latter is not therefore disproved; and that method of representation which harmonises with it may be the most true to nature, and for my own part, considering the relation of a vacuum to the magnetic force and the general character of magnetic phenomena external to the magnet, I am inclined to the notion that in the transmission of the force there is such an action, external to the magnet, than that the effects are merely attraction and repulsion at a distance. Such an action may be a function of the aether; for it is not at all unlikely that, if there be an aether, it should have other uses than simply the conveyance of radiations.

Faraday's ideas about lines of force did not, at first, make a great appeal to physicists brought up on the theory of attractions based upon action at a distance according to a law of inverse squares. For example, Sir George Airy in a letter to the Rev. J. Barlow which was passed to Faraday for comment, wrote:

... The effect of a magnet upon another magnet may be represented *perfectly* by supposing that certain parts act just as if they pulled by a string, and that other parts act just as if they pushed with a stick. And the representation is not vague, but is a matter of strict numerical calculation; and when this calculation is made on the simple law of the inverse square of distance, it does (numerically) represent the phenomena with precision. I can answer for this, because we are perpetually making this calculation. I know the difficulty of predicting the effects of evidence on other people's minds but I declare that I can hardly imagine anyone who practically and numerically knows this agreement, to hesitate an instant in the choice between this simple and precise action, on the one hand, and anything so vague and varying as lines of force, on the other hand.

You know the French mathematicians have calculated the effect of induction on the same laws.†

By this time, however, Faraday was not altogether without mathematical champions. He wrote in his reply to Barlow:

My dear Barlow—I return you Airy's second note. I think he must be involved in some mystery about my views and papers: at all events, his notes mystify me. In the first, he splits the question into (a) action inversely as the square of the distance, and (b) metaphysics. What the first has to do with my consideration I cannot make out. I do not deny the law of action referred to in all like cases; nor is there any difference as to the mathematical results (at least if I understand Thomson and Von Riess) whether he takes the results according to my view or that of the French mathematicians. Why, then, talk about the inverse square of the distance? I had to warn my audience against the sound of this law and its supposed opposition on my Friday evening, and Airy's note shows that the warning was needful.‡

As the theory left Faraday's hands it is difficult not to sympathize with Airy's complaint of lack of precision. Faraday also appears to have had difficulty over the law of inverse squares in gravitation. Maxwell wrote to him saying "I, for my part, cannot realize your dissatisfaction with the law of gravitation, provided you conceive it according to your own principles"—words which Faraday said gave him great comfort. His theory, however, needed the precision which mathematics alone could give it, to make it useful, and this he could not provide himself. Fortunately his cause was championed by Maxwell who undertook this task and when it left Maxwell's hands it had been completely transformed.

Of Faraday, Tyndall wrote:

... in his search for the unity of all forces he made all his great discoveries.

Later in life a new image of matter came into his mind. He immaterialised matter into "centres of force" and he materialised the directions in which matter tends to move into "physical lines of force". What he took from matter at its centre and gave to force, he partly gave back to matter in the lines of its motion. By this he enlarged and added to

† Bence Jones, p. 349.

‡ Bence Jones, p. 350.

subjects which he thought naturally possible for experiment to attack, and to experiment he went to test his ideas, and though he failed to realise his imaginations, yet, by his genius and truthfulness, and handicraft, he filled his experimental researches with new and connected facts, and thus he left to science a monument of himself which will last in all its grandeur for ages.

CHAPTER VI

# The Life of James Clerk Maxwell

THE work and ideas of Faraday were developed and given mathematical form by James Clerk Maxwell, whose name will always remain associated with the equations of the electromagnetic field. He was the son of John Clerk Maxwell and his wife Frances who, before her marriage, was a Miss Cay, from N. Charlton, Northumberland. They were upper middle class and both have been described as of practical bent, the husband particularly being interested in the applications of science, but both families had access to scientific circles. The name Maxwell was added to that of Clerk on the inheritance of an estate that had passed into the family through a Miss Maxwell at an earlier time.

James Clerk Maxwell was born on the 13 June 1831, at 14 India Street, Edinburgh, his parents not living as yet on the estate in Kirkcudbrightshire, where there was no house for them. The estate itself was but the remnant of one much larger at Middlebie, in the neighbouring county of Dumfriesshire, which had been cut up and sold to cover debts before John Clerk Maxwell inherited it. He, himself, built the house at Glenlair, where James spent most of his childhood and James was thus brought up as a countryman rather than a townsman. When James was 8 years old his mother died at the age of 47, of the same disease from which he, himself, was later to die at a very similar age. His early formal education was at the hands of a private tutor who, however, does not seem to have been very successful, James being reported as slow at learning. At the age of 10 he was taken to live with his paternal aunt, Mrs. Wedderburn, at 13 Heriot Row, Edinburgh, and was sent to school at Edinburgh Academy. He joined a class in the second month of their second year; his country clothes and speech appeared odd and he was badly teased. He was given the nickname "Dafty", which stuck to him for a long time, and he was for much of his school days the odd man out. These experiences may go a long way to explain his early habit of retiring within himself and his manner of speaking which was often difficult to follow.

Nevertheless, James's boyhood was not passed without considerable advantages. He spent occasional holidays at Professor Thomson's house (Lord Kelvin) and after James was about 14 and had shown some promise at school by winning prizes for mathematics, his father frequently took him to meetings of the Edinburgh Society of Arts and the Royal Society of Edinburgh. Both father and son were in close touch with Professor Forbes at the University and the latter secured the publication of James's first scientific paper, on the drawing of certain oval curves, written while he was still at school. Forbes and Maxwell commenced, at this stage, a lifelong friendship and it was Forbes's encouragement, by allowing Maxwell to use his apparatus when an undergraduate, that did much to foster and develop his scientific interests. Maxwell was also taken by his uncle, John Cay, to see Nicol—a friend of David Brewster—from whom he received, a short time later, a present of one of the well-known polarizing prisms. A year below Maxwell at school was P. G. Tait, and the two became very intimate and used to exchange manuscripts on mathematical topics. At the age of 16 Maxwell entered Edinburgh University, where he remained for three years. His holidays were mostly spent at Glenlair, where he had time to develop his interests in polarized light, electricity and geometry. He had a geometrical paper read at the Royal Society of Edinburgh in 1849 and another on the Equilibrium of Elastic Solids in 1850. In brief, in his early time Maxwell became acquainted with a very distinguished circle of Scottish scientists, and by his life in the country and the opportunities of leisure which were his, he made intimate contact with nature and natural phenomena.

Maxwell went up to Peterhouse, Cambridge, in the autumn of

1850, at the age of 19, the College to which P. G. Tait had already gone two years earlier. The number of able mathematicians at the College, however, was such that the prospect of obtaining a fellowship there seemed small and at the end of the first term, on his father's advice, he migrated to Trinity, which many of his friends in Edinburgh had advised from the start. For the first year he lived in lodgings in King's Parade but afterwards he had rooms in college—G.6 Great Court in 1852 and 1853, and I.4 Nevile's Court in 1854 and 1855. In 1854 he passed the Mathematical Tripos with the position of Second Wrangler. Routh (of Peterhouse and "Routh's Rule") was First Wrangler but shared the Smith's Prize with Maxwell.

Maxwell was athletic, but sport in the ordinary sense made little appeal to him. He was short-sighted. His love of animals made shooting on the Scottish moors abhorrent to him. His principal means of taking exercise were walking, swimming, sculling and riding. On one occasion, after staying with friends in the Lake District, he walked home to Glenlair from Carlisle, a distance of about 50 miles. Of his swimming P. G. Tait said: "He used to go up on the pollard at the bathing shed, throw himself *flat on his face* in the water, dive and cross, then ascend the pollard on the other side, project himself *flat on his back* in the water. He said it stimulated the circulation!" He was a skilful horseman and, when he was older, both he and his wife enjoyed riding. For many years his wife rode a pony called "Charlie", which Maxwell had broken in himself, riding side-saddle with a carpet to simulate a lady's riding habit. He rode a black horse called "Dizzy" which was said to have been the despair of previous owners. He also took exercise in the gymnasium and was interested in the science of movement. The latter no doubt accounted for his keenness on the "diabolo" or the "devil on two sticks" as he called it.

Maxwell also possessed a sense of quiet humour which is attested to by many who knew him. Two examples of practical jokes by him are recorded in his biography by Campbell and Garnett. When he came up for his M.A. degree in 1857 he brought with him his "Dynamical Top" and exhibited it at a tea party in his rooms. His friends left when it was still spinning. Next morning Maxwell spotted one of them coming across the court. He leapt out of bed, started the top spinning and quickly got back in again. Much later, when Cavendish Professor he came across some thick fur in a boiler in which the scale appeared to him to be a bit oolitic. He sent a specimen to the Professor of Geology with a request that he should identify the deposit. This was duly done—and quite correctly!

He read in the Classics, which had been the staple of his early education, though it was a year and a half after entering Cambridge that he passed his Little-go. He took considerable interest in philosophy, which he had studied at Edinburgh University, particularly the theory of knowledge. He wrote a considerable quantity of English verse. He was by temperament conservative and religious, and he tended to support the ecclesiastical line in the debates with the evolutionists which were in full swing in his time. His principal argument, that the atoms of which matter is made could not have evolved, would not be acceptable today—at least it would require to be pushed a stage further back. It serves chiefly as a reminder of the mutability of scientific opinion. Maxwell, however, confessed that he had "no nose for heresy" and had friends on both sides of the controversy. He obtained his fellowship at Trinity in October 1855, at his second attempt. It was awarded on examination and he was one of three mathematicians selected. He immediately started to lecture on optics and hydrostatics.

Throughout his life Maxwell remained powerfully attracted to his country home at Glenlair. This may well have been due in some measure to the awkwardness of his first emergence into the world at school, and the comparative slowness with which he overcame the obstacles. Since his mother died he had formed a great attachment to his father who then lived alone, and when the professorship of Natural Philosophy became vacant at Marischal College, Aberdeen, he applied for the post. The long summer vacation would enable him to live for nearly half the year at Glenlair, largely uninterruptedly. His object in being near his father, however, was never attained for just before he returned to Cambridge for his last summer term his father died. Maxwell returned to Cambridge in the usual way and his candidature at Marischal College went forward, and he was later appointed. Glenlair remained his home.

He started at Aberdeen in November 1856. The work of a professor in a Scottish university at that time was more like that of a schoolmaster. Boys entered at an early age and their studies were very elementary. There has been a good deal of debate about Maxwell's ability as a teacher. At Aberdeen his classes were large—some seventy or eighty strong—his manner was curious, often dreaming and suddenly awakening, and he was probably over the heads of most of his pupils. There is little doubt that a few of the best profited greatly from his instruction but the rest were left behind. While at Aberdeen he married the daughter of the Principal of the College—Frances Mary Dewar—and this alone would seem to indicate that he is unlikely to have been looked upon in any way as a misfit.

After 4 years, Marischal College was united with King's College to form Aberdeen University, and the post of Professor of Natural Philosophy was given to David Thomson, the holder of the post at King's College and Maxwell's senior. Maxwell then first applied for the professorship at Edinburgh which Forbes had just resigned, but it was given to another of his friends, P. G. Tait. In the summer of 1860 he was appointed to King's College, London.

It was while he was at King's College, London, that his most important electrical work was published or initiated. He held the post for 5 years. He and his wife lived for most of that time at 8 Palace Gardens Terrace, Kensington, and the duties of the professorship kept them in London for 9 months of the year. Mrs. Maxwell helped her husband in some of his experiments on colour vision and they had a laboratory in the attic which ran the whole length of the house. They must have longed, however, to make fuller use of their home in the country at Glenlair.

In London, Maxwell continued his practice, which he had started in Cambridge and followed at Aberdeen, of giving evening

*James Clerk Maxwell and his wife*
From a portrait in the Cavendish Laboratory, Cambridge.

Glenlair: Maxwell's house in Kirkcudbrightshire. Set among the pleasant rolling country of the lower Galloway hills it is hardly surprising that it tended to attract him and his wife away from the cares of academic life. In addition to the mansion, which was gutted by fire in September 1929 and is now a ruin, Maxwell owned the following farms: Nether and Upper Glenlair, Nether Corsock, Hillside, Blackhill and Little Mockrum.

Glenlair is a Gaelic word meaning middle of the glen or valley.

lectures to working men. One of the opportunities which residence in London afforded was the possibility of getting to know Faraday, then living at Hampton Court, though still continuing with his duties at the Royal Institution. Maxwell gave his first lecture there on 17 May 1861 "On the Theory of the Three Primary Colours". At other times he was among the audience. On one occasion, after he had published his first paper on the kinetic theory of gases, he became wedged in a crowd attempting to get out and was spotted by Faraday who called out "Ho, Maxwell, cannot you get out? If any man can find his way through a crowd it should be you!" An important characteristic feature in their reasoning, which Faraday and Maxwell shared, was the habit of thinking in terms of physical pictures. Although an accomplished mathematician, Maxwell, as has been remarked by many, relied more on diagrams and geometrical notions than on symbols. For example, in his paper of 20 April 1857 "On a Dynamical Top" published in the *Transactions of the Royal Society of Edinburgh*, he wrote:

> The mathematical difficulties of the theory of rotation arise chiefly from want of geometrical illustrations and sensible images, by which we might fix the results of analysis in our minds. . . . Even the velocity of rotation about the axis requires a careful definition and the proposition that, in all motion about a fixed point, there is always one line of particles forming an instantaneous axis, is usually given in the form of a very repulsive mass of calculation. Most of these difficulties may be got rid of by devoting a little attention to the mechanics and geometry of the problem before entering on the discussion of the equations.

During his London period there appeared in 1860 his first paper on the kinetic theory of gases and his second on colour analysis. In 1861 he published his paper "On Physical Lines of Force" which we discuss in the next chapter. (He had previously written "On Faraday's Lines of Force" as he was starting at Aberdeen, the work for which must have been done while he was at Cambridge.) Three years later came his principal paper on electricity, "A Dynamical Theory of the Electromagnetic Field". In his last year at King's College he must have been working on the dynamical theory of gases. Two papers from his pen on this subject appeared in the following year. It has been suggested that the origins of his ideas in this field lay in the statistical theory he had employed in dealing with the problem of Saturn's rings, for which he had been awarded the Adam's Prize of 1857. Maxwell resigned the chair at King's College in the spring of 1860. He was succeeded by W. G. Adams, but continued to lecture to his working men during the following winter.

The beginning and end of the period at King's College were both marked by serious attacks of illness. In September 1860 he contracted smallpox at Glenlair—supposedly at the fair at which "Charlie" was bought. In September 1865, also at Glenlair, he suffered from a severe attack of erysipelas. On both occasions his wife was his nurse, a profession in which he, himself, was not inexperienced. While at Cambridge he had nursed a friend, one Pomeroy who died later in the Indian Mutiny, through a serious illness; he had personally attended his father at Glenlair; while in London he had his brother-in-law at his house while he underwent a painful operation, and he also looked after his wife personally when she was ill.

After London, Maxwell lived at Glenlair, though frequent visits were made to London, and to Cambridge where he acted either as Moderator or Examiner in the Mathematics Tripos. In 1867 he and his wife made a tour of Italy. The time was mainly spent, however, in quiet country life and in giving the final form to his electrical theory by the writing of his Treatise. While in London he had continued to take much interest in his country property. In 1860 some peacocks were purchased from a Mrs. McCunn of Ardhallow and they usually sat and adorned the front porch. At a later date Lord Kelvin became interested in having some of the birds, and Maxwell wrote to him saying that when he first acquired them, the garden at Ardhallow was the finest on the coast. When Mr. McCunn died his wife let the gardener be master and the peacocks were destroyed, and the garden ceased to produce the commonest vegetables and became quite a desert! Maxwell was rarely seen walking without a dog. Later on at Cambridge they went with him to the Cavendish Laboratory where, on one occasion, one of them sat on an insulating stool and allowed itself to be charged electrically by being rubbed with a cat's skin. A cat's skin is charged positively when rubbed on other materials but on this occasion it was the dog that received the positive charge. "A live dog is better than a dead lion" said Maxwell.

In 1871 he was invited to become the first Cavendish Professor of Physics at Cambridge and with some reluctance he finally accepted. He gave his inaugural lecture in October and his first duty was the design and equipment of the Cavendish Laboratory, the funds for which had just been given to the University by its Chancellor, the Duke of Devonshire, after whom it is named. Apart from this, Maxwell's first two years in this post saw the finishing touches put to the *Treatise on Electricity and Magnetism* and in supervising its progress through the press. It was published in 1873, being preceded in 1871 by the less well-known *Treatise on Heat*. After that Maxwell devoted what time he had to dispose of to the editing of *The Electrical Researches of the Hon. Henry Cavendish* and the result was published in 1879. In that year he also started on a new edition of the electrical treatise but his health gave way. He went back to Glenlair for the summer, suspecting only a temporary lack of fitness, and in the hope that the bracing air would help him to recover, instead of which his illness became worse and he was told on 2 October by his physician that he had but one month to live. He was brought back to Cambridge in order to have better medical supervision, but he died on 8 November at the age of 48.

In addition to his major contributions to science with which his name will always be associated, the Cavendish Laboratory in Cambridge provides a striking memorial to James Clerk Maxwell. Its design was his and, though he was supported by able assistants, it was he who laid the foundations for the school of research which brought world-wide fame to the institution before the century was out. Shortly after Maxwell died Lord Kelvin wrote:

> The influence of Maxwell at Cambridge had undoubtedly a great effect in directing mathematical studies into more fruitful channels

> than those in which they had been running for many years. His published scientific papers and books, his action as an examiner at Cambridge, and his professorial lectures, all contributed to this effect; but above all, his work in planning and carrying out the arrangements of the Cavendish Laboratory. There is, indeed, nothing short of a revival of Physical Science at Cambridge within the last fifteen years, and this is largely due to Maxwell's influence.

Maxwell's last public lecture was the Rede Lecture "On the Telephone", which he delivered in Cambridge on 24 May 1878. In it he said:

> In a university we are especially bound to recognise not only the unity of Science itself, but the communion of the workers of science. We are too apt to suppose that we are congregated here merely to be within reach of certain appliances of study, such as museums and laboratories, libraries and lectures, so that each of us may study what he prefers. I suppose that when the bees crowd round the flowers it is for the honey that they do so, never thinking that it is the dust which they are carrying from flower to flower which is to render possible a more splendid array of flowers and a busier crowd of bees in the years to come.

CHAPTER VII

# Maxwell's Development of Electromagnetic Theory

WHILE Faraday worked in a highly idiosyncratic way developing his own ideas in the light of experimental enquiries which he made largely his own, Maxwell sought to translate these same ideas into the language already employed by science. In his *Treatise on Electricity and Magnetism* (p. 175) he compared the work of Ampère, an outstanding mathematical physicist, with that of Faraday, the experimentalist.

> The method of Ampère . . . [he said] though cast into an inductive form, does not allow one to trace the formation of the ideas which guided it. We can scarcely believe that Ampère really discovered the law of action by means of the experiments which he describes. We are led to suspect, what indeed he tells us himself, that he discovered the law by some process which he has not shown us, and that when he had afterwards built up a perfect demonstration he removed all traces of the scaffolding by which he had raised it.
>
> Faraday, on the other hand, shows us his unsuccessful as well as his successful experiments, and his crude ideas as well as his developed ones, and the reader, however inferior to him in inductive power, feels sympathy even more than admiration, and is tempted to believe that, if he had the opportunity, he too would be a discoverer. Every student should therefore read Ampère's research as a splendid example of scientific style in the statement of a discovery, but he should also study Faraday for the cultivation of a scientific spirit, by means of the action and reaction which will take place between the newly discovered facts as introduced to him by Faraday and the nascent ideas in his own mind.

He goes on to say:

> It was perhaps for the advantage of science that Faraday, though thoroughly conscious of the fundamental forms of space, time and force, was not a professed mathematician. He was not tempted to enter into the many interesting researches in pure mathematics which his discoveries

would have suggested if they had been exhibited in a mathematical form, and he did not feel called upon either to force his results into a shape acceptable to the mathematical taste of the time, or to express them in a form which mathematicians might attack. He was left at leisure to do his proper work, to coordinate his ideas with his facts, and to express them in natural untechnical language.

It is mainly in the hope of making these ideas the basis of a mathematical method that I have undertaken this treatise.

In this task Maxwell was brilliantly successful, but to regard his contribution as solely one of translation would grossly undervalue it. It is true that the fundamental concepts of Maxwell's theory are to be found in Faraday, who himself got very close to an electromagnetic theory of light in his "Thoughts on Ray Vibrations", but it was extremely vague and the matter could not be clinched by an estimation of the velocity of propagation. In Maxwell's hands the ideas were clarified and the whole theory given a precision which lay beyond the powers of Faraday. Maxwell's written account of the theory is also interspersed with numerous interesting comments, of which the quotation given above is an example, which make his writing stimulating to read.

Maxwell's contributions to electromagnetic theory were made in four stages. The first appeared in a long paper which was read to the Cambridge Philosophical Society on 10 December 1855 and on 11 February 1856. It was entited "On Faraday's Lines of Force". The second was made in a paper of similar length, published in parts, in the *Philosophical Magazine* (vol. XXI) in March 1861. The third is in the paper which has been selected for reproduction in this book and was called "A Dynamical Theory of the Electromagnetic Field". It was received by the Royal Society on 27 October 1864 and published in volume CLV of the *Transactions*. Maxwell made his fourth contribution by writing his *Treatise on Electricity and Magnetism* which, as its name implies, was a more comprehensive account of the whole subject from the point of view of the field theory. It was first published in 1873. We consider first the paper "On Faraday's Lines of Force" which illustrates the beginning of the attack which Maxwell made on the problem.

In this paper he took as his object simply the expression of

Faraday's ideas of lines of force in mathematical language. He advanced no physical theory to account for them and, though the mathematical symbols are based on a physical analogy, they are purely descriptive of the properties of the lines of force. In one of his characteristic incidental remarks he says:

> The first process therefore in the effectual study of the science, must be one of simplification and reduction of the results of previous investigations to a form in which the mind can grasp them. The results of this simplification may take the form of a purely mathematical formula or of a physical hypothesis. In the first case we entirely lose sight of the phenomena to be explained, and though we may trace out the consequences of given laws, we can never obtain more extended views of the connexions of the subject. If, on the other hand, we adopt a physical hypothesis, we see the phenomena only through a medium, and are liable to that blindness to facts and rashness in assumption which a partial explanation encourages. We must therefore discover some method of investigation which allows the mind at every step to lay hold of a clear physical conception, without being committed to any theory founded on physical science from which that conception is borrowed, so that it is neither drawn aside from the subject in pursuit of analytical subtleties, nor carried beyond the truth by a favourite hypothesis.
>
> In order to obtain physical ideas without adopting a physical theory, we must make ourselves familiar with the existence of physical analogies. By a physical analogy I mean that partial similarity between the laws of one science and those of another, which makes each of them illustrate the other. . . .†

Two examples of such analogies given by Maxwell are first that between the refraction of light and the deflection of a particle by an intense but narrow accelerating field, and second, that between the flow of heat and attraction by action at a distance. The latter analogy had recently been developed by William Thomson in the *Cambridge Mathematical Journal.*

For the purpose of describing the properties of lines of force Maxwell first develops a theory of the motion of an incompressible fluid.

> It is not even a hypothetical fluid introduced to explain the phenomena. It is merely a collection of imaginery properties which may be employed for establishing certain theorems in pure mathematics in a way more intelligible to many minds and more applicable to physical problems than that in which algebraic symbols alone are used.

† *Collected Papers,* I, p. 155.

The object is to construct a mathematical calculus to which the lines of force can be attached and the fluid merely helps the intellect in this task. The fluid may be as irrelevant to magnetism and electricity as the flow of heat is to action at a distance.

The analogy to lines of force which Maxwell constructs is that of the uniform motion of an imponderable and incompressible fluid through a resisting medium. The fluid itself is uniform, it possesses no inertia and its motion is resisted by a force proportional to its velocity. It will move, therefore, only when subjected to pressures. Surfaces of equal pressure are discussed. They are shown to be at right angles to the lines of fluid motion. The fluid flow is mapped by lines and tubes of flow and the whole of space divided into cells defined by the parts of unit tubes cut off by neighbouring surfaces of equal pressure for which $\delta p = 1$. Writing the resistance $R = ku$, where $u$ is the velocity of the fluid and $k$ a constant of proportionality, the pressure at a distance $r$ from a source of strength $S$ becomes

$$p = \frac{kS}{4\pi r}.$$

If $p$ and $S$ refer to one source and $p'$ and $S'$ to another while $r$ is the distance between them,

$$S'p = \frac{kSS'}{4\pi r} = Sp'.$$

A number of other theorems with the electrostatical counterparts of which the reader will be familiar are then proved. On this basis it is possible to state the analogy existing between the flow of the fluid and electricity. Maxwell identifies electric potential $V$ with $-p$, force with $-dp/dr = ku$ and electrical charge with $kS/4\pi$.

The theory of dielectrics depends upon the analogy between a change in dielectric constant and a change in resistance to fluid flow. Sources and sinks can be suitably distributed over surfaces so as to simulate in a uniform space any system of existing lines of force. The sources and sinks correspond to induced charges.

The analogy of fluid flow is equally applicable to magnetism and in this paper it is applied to permanent magnets, and to paramagnetic and diamagnetic induction (on the lines of Faraday's theory of conduction of lines of force). Another and equally independent application of the analogy is to the flow of electric currents familiar to every elementary schoolboy.

The analogy is simply used to portray the general properties of lines of force. The use of the lines in determining the action between two circuits is then discussed on the basis of the equivalent magnetic shell, but perhaps the most interesting part of this paper is that which deals with the induction of currents and Faraday's electro-tonic state. The facts of electromagnetic induction and their connection with the lines of force are taken as Faraday presented them. The number of lines linking a circuit, however, is calculated by means of a vector potential integrated round the length of the circuit and this Maxwell identifies as the measure of the electrotonic state. However, like Faraday, he is unable to discover any property of the conductors which corresponds, by which the state can be recognized, and he thus has to rest content with symbolism. Of Faraday's concept he wrote:

> The conjecture of a philosopher so familiar with nature may sometimes be more pregnant with truth than the best established experimental law discovered by empirical enquiries, and though not bound to admit it as a physical truth, we may accept it as a new idea by which our mathematical conceptions may be rendered clearer.
>
> In this outline of Faraday's electrical theories, as they appear from a mathematical point of view, I can do no more than simply state the mathematical methods by which I believe that electrical phenomena can be best comprehended and reduced to calculation, and my aim has been to present the mathematical ideas to the mind in an embodied form, as systems of lines or surfaces, and not as mere symbols, which neither convey the same ideas nor readily adapt themselves to the phenomena to be explained. The idea of the electro-tonic state, however, has not yet presented itself to my mind in such a form that its nature and properties may be clearly explained without reference to mere symbols. . . .†

Maxwell does not employ vector methods and his papers tend, as a result, to become rather voluminous. His notation in this paper is briefly as follows. The vector $\alpha_0$, $\beta_0$, $\gamma_0$ is the vector

† *Collected Papers*, I, p. 187.

potential; $\alpha_1, \beta_1, \gamma_1$ is the magnetic intensity; $\alpha_2, \beta_2, \gamma_2$ is the electric intensity; $a_1, b_1, c_1$ is the magnetic flux and $a_2, b_2, c_2$ the electric current. The paper contains a deduction of the law of the induction of currents on the lines of Helmholtz's calculation. It is as follows, using a vector notation and taking the first of Maxwell's symbols to represent the whole vector.

The rate of dissipation of electrical energy is $\int (a_2 \cdot \alpha_2)\, d\tau$.

The rate of working against external forces is

$$\frac{d}{dt}\int \frac{(a_2\alpha_0)}{4\pi}\, d\tau.$$

Whence

$$(a_2 \cdot \alpha_2) + \frac{(a_2 \cdot d\alpha_0)}{4\pi dt} = 0$$

and

$$\alpha_2 = -\frac{1}{4\pi}\frac{d\alpha_0}{dt}.$$

This proof is only satisfactory if $\alpha_0$ arises from a permanent magnet, because if a contribution is made towards it by a second circuit then the same quantity of external work would be performed, assuming as before that the currents are maintained constant, but energy would appear not only in the first circuit but in the second also, the extra amount dissipated coming from that stored in the field.

The paper finishes with twelve examples on the use of the theory. What it provided was the basic mathematical apparatus with which electrical problems could be approached. As Maxwell himself remarked: "I do not think it contains even the shadow of a true physical theory; in fact its chief merit as a temporary instrument of research is that it does not, even in appearance, *account* for anything".†

The mathematical calculus described in this paper is, of course, as appropriate to a theory of action at a distance as it is to one of

† *Collected Papers*, I, p. 207.

lines of force. Indeed it was one of Maxwell's purposes to show that the two theories led to the same results, a fact which was by no means generally appreciated, as we have seen. By his introduction of the vector potential as a measure of the electro-tonic state, however, Maxwell actually makes it easier to dispense with lines of force, even in the case of induction of currents, the laws of which had been enunciated by Faraday in terms of these lines. The electromotive force induced into each element of a circuit becomes simply the rate of change of the vector potential at the position of the element and the vector potential is calculable directly from the currents which produce the magnetic field.

There is no mention in this paper of that feature which became central in Maxwell's later work, namely the displacement current. For this reason there is no reference to the possibility of radiation. Apart from the briefest reference to the possibility of treating electrostatic fields by the method of flux, little attention is given to static electricity. It was only when a so-called physical theory of lines of force had been adumbrated that a concept such as the displacement current became possible. The origin of this can be traced directly to Faraday's views concerning electrostatic induction. This concept was developed by Maxwell in the papers published in the *Philosophical Magazine* just over 5 years later, in which was produced the theory of "Physical Lines of Force"—that is to say, lines of force endowed with some structural existence as distinct from mathematical concepts used only for purposes of calculation. How far mental concepts can be endowed with physical reality we will not discuss here.

Faraday, it will be remembered, wished to explain electrostatic induction by polarization in the intervening medium, the dielectric. This he considered to arise from the action of contiguous particles. To those who thought of a vacuum as empty space, the possibility of induction through a vacuum cast serious doubts on this view and, as we have seen, Faraday was not very convincing in dealing with them. Contiguous particles could be inches apart, which could hardly be looked upon as action at insensible distances, which was Faraday's basic idea. However, neither Faraday nor Maxwell really regarded a vacuum as mere empty space, Faraday at first thinking of it as filled with ordinary ponderable matter and Maxwell employing an aether instead and thus agreeing with Professor Hare. Even in his third paper—the one reproduced in this book—in which the electromagnetic theory was fully developed, Maxwell says:

> The electromagnetic field is that part of space which contains and surrounds bodies in electric or magnetic conditions.
>
> It may be filled with any kind of matter, or we may endeavour to render it empty of all gross matter, as in the case of Geissler's tubes and other so-called vacua.
>
> There is always, however, enough of matter left to receive and transmit the undulations of light and heat, and it is because the transmission of these radiations is not greatly altered when transparent bodies of measurable density are substituted for the so-called vacuum, that we are obliged to admit that the undulations are those of an aethereal substance, and not of gross matter, the presence of which merely modifies in some way the motion of the aether.
>
> We have, therefore, some reason to believe, from the phenomena of light and heat, that there is an aethereal medium filling space and permeating matter, capable of being set in motion and of transmitting that motion to gross matter, so as to heat it and affect it in various ways. . . . Professor Thomson has argued that the medium must have a density capable of comparison with that of gross matter and has even assigned an inferior limit to that density.
>
> We may, therefore, receive as a datum derived from a branch of science independent of that with which we have to deal, the existence of a pervading medium, of small but real density, capable of being set in motion and of transmitting motion from one part to another with great but not infinite velocity.

Still later, in the *Treatise* of 1873, which contains his last words on the subject, he continues to speak very definitely of a medium. Here he says (vol. II, p. 470):

> A theory of molecular vortices, which I worked out at considerable length, was published in the *Phil. Mag.* . . .
>
> I think we have good evidence for the opinion that some phenomenon of rotation is going on in the magnetic field, that this rotation is performed by a great number of very small portions of matter, each rotating on its own axis, this axis being parallel to the direction of the magnetic force, and that the rotations of these different vortices are made to depend on one another by means of some kind of mechanism connecting them.
>
> The attempt which I then made to imagine a working model of this mechanism must be taken for no more than it really is, a demonstration that mechanism may be imagined capable of producing a connexion

mechanically equivalent to the actual connexion of the parts of the electromagnetic field. The problem of determining the mechanism required to establish a given species of connexion between the motion of the parts of a system always admits of an infinite number of solutions. Of these, some may be more clumsy or more complex than others, but all must satisfy the conditions of mechanism in general.

The following results of the theory, however, are of higher value—

(1) Magnetic force is the effect of the centrifugal force of the vortices.

(2) Electromagnetic induction of currents is the effect of the forces called into play when the velocity of the vortices is changing.

(3) Electromotive force arises from the stress on the connecting mechanism.

(4) Electric displacement arises from the elastic yielding of the connecting mechanism.

From the facts that electrostatic charge always resided on the surface of conductors only and that positive and negative charges could only be created together in equal amounts and never separately, we have seen that Faraday was persuaded to concentrate his attention on the surrounding medium and to look upon charges on bodies as simply the result of polarization in it. In effect the charges were reduced to the terminations of the lines of force. In his second paper on electrodynamics Maxwell elaborated the same point of view. Electricity he put firmly in the medium and it permeated all space, including that occupied by matter. In conductors the electricity was capable of being transferred from one particle to the next. In non-conductors, including space or "so-called vacua", elastic displacement or microscopic circulation were the only movements possible. It was to illustrate these properties that Maxwell constructed his model. Later Maxwell, himself, dismantled a good deal of the detail of the model and today very little indeed of it is left. The story of the development of electromagnetic theory appeared to many at the time somewhat reminiscent of the Cheshire Cat.

> "I wish you wouldn't keep appearing and vanishing so suddenly", replied Alice, "you make one quite giddy". "All right" said the Cat; and this time it vanished quite slowly beginning with the end of the tail and ending with the grin, which remained some time after the rest of it had gone. "Well! I've often seen a cat without a grin", thought Alice; "but a grin without a cat! It's the most curious thing I ever saw in my life!"

It is a question of experience

At the outset of the theory the electromagnetic aether appeared clothed in much circumstantial detail. In constructing a model a choice had first to be made between two possibilities. Either the electric current could be treated as a flow and magnetic lines of force as a tension arising from a vortical motion around them, or, alternatively, the lines of magnetic force could be treated as a flow and current as a tension. Because of the transference of ponderable matter in electrolysis and the rotation of the plane of polarization of light travelling along lines of magnetic force, Maxwell had no difficulty in selecting the first alternative. The following is an outline of how he worked out the idea in his second contribution to the subject, "On Physical Lines of Force".

Lines of magnetic force, the transmission of which Faraday had found rather baffling, run from north-seeking to south-seeking poles between which there is an attraction. In terms of a medium in which the lines of force are thought of as existing, these lines behave as though in a state of tension longitudinally. The forces observed with electric currents in magnetic fields indicated that the lines of magnetic force also tended to expand laterally. This behaviour can be imitated by a system of vortices, the centrifugal force of which causes them to expand laterally and contract longitudinally. If not prevented by a suitable distribution of forces, a medium containing vortices would expand laterally, allowing the diameter of each vortex to increase and the velocity to diminish in the same proportion. The magnetic field, according to this view, could be looked upon as a medium with a certain defect of pressure in the direction of the axes of the vortices. This pressure difference would be proportional to the square of the peripheral velocity of the vortices and Maxwell writes it $\mu v^2/4\pi$.

If the general pressure is $p_1$ and the longitudinal tension $T$ arising from the difference of pressure, it is evident from Fig. 7 that the forces on the face $yz$ are a pressure

$$p_{xx} = lTS\frac{l}{S} - p_1 = l^2T - p_1$$

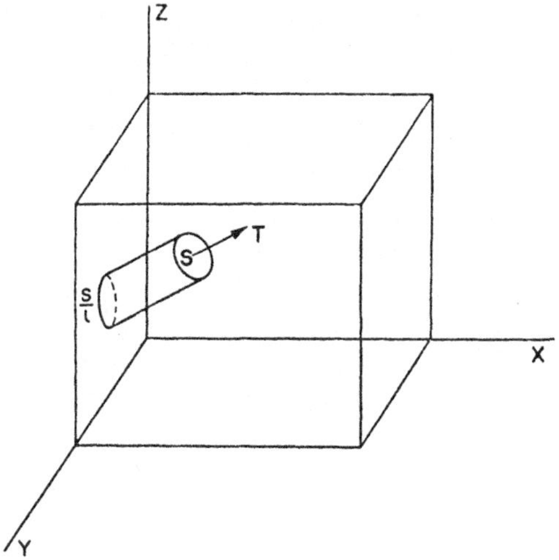


FIG. 7.

together with a tangential force in the direction of $z$ of

$$p_{zx} = nTS\frac{l}{S} = lnT$$

per unit area, and a similar force in the direction of $y$ of

$$p_{yx} = lmT$$

where $l$, $m$, $n$ are the direction cosines of the axes of the vortices.

We have

$$T = \mu v^2/4\pi$$

and writing

$$\alpha = lv, \quad \beta = mv, \quad \text{and} \quad \gamma = nv \tag{1}$$

we have

$$p_{xx} = \frac{\mu}{4\pi}\alpha^2 - p_1$$

$$p_{yx} = \frac{\mu}{4\pi}\alpha\beta$$

$$p_{zx} = \frac{\mu}{4\pi}\alpha\gamma.$$

The remaining terms may be written down by inspection. In this notation $p_{xy}$ is the force in the direction of $x$ on unit area of the plane parallel to $y = 0$. The force per unit volume in the direction of $x$ on an element of volume, arising from variations in the intensity and distribution of the vortices, will be

$$X = \frac{d}{dx} p_{xx} + \frac{d}{dy} p_{xy} + \frac{d}{dz} p_{xz}.$$

Using the above expressions for the stresses this may be written

$$X = \frac{\alpha}{4\pi}\left(\frac{d}{dx}(\mu\alpha) + \frac{d}{dy}(\mu\beta) + \frac{d}{dz}(\mu\gamma)\right) + \frac{\mu}{8\pi}\frac{d}{dx}(\alpha^2 + \beta^2 + \gamma^2)$$
$$- \frac{\mu\beta}{4\pi}\left(\frac{d\beta}{dx} - \frac{d\alpha}{dy}\right) + \frac{\mu\gamma}{4\pi}\left(\frac{d\alpha}{dz} - \frac{d\gamma}{dx}\right) - \frac{dp_1}{dx}.$$

The vector $(\alpha \,.\, \beta \,.\, \gamma)$ is now identified with the magnetic intensity, and writing for conciseness $\boldsymbol{\alpha}$ to represent this vector and similarly $\boldsymbol{X}$ for the vector $X$, $Y$, $Z$, we see that the first bracket is

$$\boldsymbol{X} = \frac{\boldsymbol{\alpha}}{4\pi} \operatorname{div} \mu\boldsymbol{\alpha} = \boldsymbol{\alpha} m$$

where $m$ is the density of "magnetic matter".

The second term,

$$\boldsymbol{X} = \frac{\mu}{8\pi} \operatorname{grad} \alpha^2$$

shows that bodies for which $\mu$ is positive move into the strongest part of the field.

The third and fourth terms are together,

$$\boldsymbol{X} = \frac{\mu\boldsymbol{\alpha}}{4\pi} \times \operatorname{curl} \boldsymbol{\alpha}.$$

Now curl $\boldsymbol{\alpha}$ is equal to the current density $\boldsymbol{p}(p \,.\, q \,.\, r)$ so that these terms become

$$\boldsymbol{X} = \frac{\mu\boldsymbol{\alpha}}{4\pi} \times \boldsymbol{p}$$

and give the force on a current in a field the magnetic induction in which is $\mu\alpha$.

The last term $dp_1/dx$, merely indicates that the element will be urged in the direction in which the hydrostatic pressure diminishes.

Maxwell also gives a qualitative picture of the working of his theory. First the behaviour of a magnetic pole placed in a field of force: Suppose $N$ is a north-seeking pole in a field, the direction of the lines of force of which is that of the arrows pointing to the right in Fig. 8. The lines of force arising from the pole itself

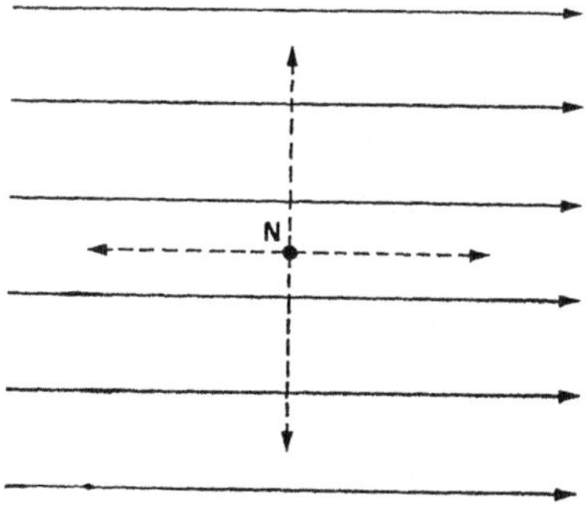


FIG. 8.

will be in the same direction as the field in the region towards the right and in the opposite direction towards the left. The field vortices will therefore be speeded up on the right and slowed down on the left. The tension will thus be greater on the right than on the left, so that the pole will be urged in the direction of the field. In a similar manner a south-seeking pole will be urged in the opposite direction.

In the case of an electric current flowing upwards in a horizontal magnetic field, as in Fig. 9, the lines arising from the current augment the field on the right and diminish it on the left. The lateral pressure is therefore greater on the right than on the left and the conductor is urged towards the left.

In the absence of current curl $\boldsymbol{\alpha} = 0$ and it is possible to write $\boldsymbol{\alpha} = \text{grad}\, \phi$, where $\phi$ is some scalar.

Thus

$$m = \frac{\mu}{4\pi} \operatorname{div} \alpha = \frac{\mu}{4\pi} \nabla^2 \phi.$$

The solution of this equation is

$$\phi = \Sigma \frac{m}{\mu r}$$

so that, as Maxwell says:

> The lines of force in a part of space where $\mu$ is uniform and where there are no electric currents, must be such as would result from the theory of "imaginery matter" acting at a distance. The assumptions of that theory are unlike those of ours but the results are identical.

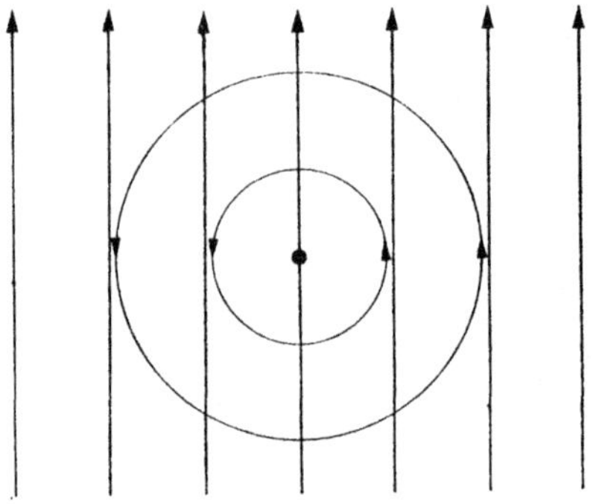

FIG. 9.

The next questions to arise are "How are these vortices set in rotation?" and "Why are they arranged according to the known laws of lines of force about magnets and currents?" Maxwell regards these questions as of a higher order of difficulty than those he has dealt with so far, and the suggestions he makes towards their solution as providing a provisional answer only.

This provisional answer is apt to strike the modern reader as very curious. In a uniform magnetic field all the vortices will rotate in the same direction so that their surfaces of contact will be moving in opposite directions. To overcome this difficulty Maxwell, adapting an idea of Johann Bernoulli the younger, put "ball bearings" between the vortices in the form of particles which rolled without slipping on the surfaces of the vortices, and it was of these particles that electricity was composed. The object of such a model was, of course, not so much to describe the actual structure of the aether as to devise a mechanical system which obeyed the same laws. The belief that some form of mechanical structure existed seems to have been accepted. It was by no means the most fantastic of such models to be put forward. For the elastic solid aether Lord Kelvin, in 1889, imagined a structure

> formed of spheres, each sphere being in the centre of a tetrahedron formed by its four nearest neighbours. Let each sphere be joined to these four neighbours by rigid bars, which have spherical caps at their ends so as to slide freely on the spheres. Such a structure would, for small deformations, behave like an incompressible fluid. Now attach to each bar a pair of gyroscopically mounted flywheels, rotating with equal and opposite angular velocities and having their axes in the line of the bar; a bar thus equipped will require a couple to hold it at rest in any position inclined to its original position and the structure as a whole will possess that kind of quasi-elasticity which was first imagined by MacCullagh.

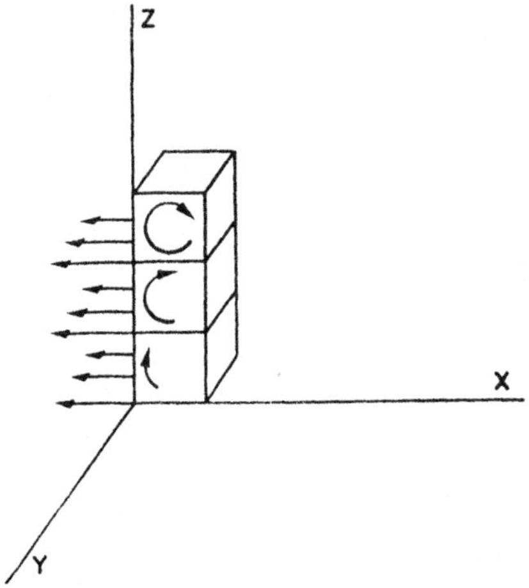


FIG. 10.

On the basis of the model with the electrical ball bearings we can calculate the transfer of particles across unit volume, caused by variations in the vorticity. To save space let us simplify the deduction by assuming cubical vortices with faces parallel to the coordinate planes. Let $\beta$ be the peripheral velocity caused by circulation about an axis in the direction of $y$. The electrical particles between the vortices will travel with a velocity equal to half the numerical difference between the peripheral velocities of neighbouring vortices. If there are $\rho$ particles per unit surface of the vortical cells, then the number crossing unit area at right angles to the axis of $x$ will be

$$-\frac{\rho}{2}\frac{d\beta}{dz}$$

Adding $\rho/2\ (d\gamma/dy)$ for a circulation about the axis of $z$, we have for the $x$ component of the particle flow,

$$p = \frac{\rho}{2}\left(\frac{d\gamma}{dy} - \frac{d\beta}{dz}\right)$$

or using vectors, $\boldsymbol{p} = \rho/2 \text{ curl } \boldsymbol{\alpha}$.

This will agree with the first law of circulation round a current $p$ as deduced from Ampère's law of action of current elements if we put

$$\rho = \frac{1}{2\pi}. \tag{2}$$

To account for a vortex being able to set its neighbours in rotation, which is necessary if the model is to account for electromagnetic induction, we have to assume that the motion of the particles is resisted.

> Now [says Maxwell] let us suppose the vortices arranged in a medium in any arbitrary manner. The quantities
>
> $$\frac{d\gamma}{dy} - \frac{d\beta}{dz} \text{ etc.,}$$
>
> will then in general have values, so that there will, at first, be electrical currents in the medium. These will be opposed by the electrical resistance of the medium; so that unless they are kept up by a continuous supply of force, they will quickly disappear and we shall have
>
> $$\frac{d\gamma}{dy} - \frac{d\beta}{dz} = 0 \text{ etc.;}$$
>
> that is $\alpha dx + \beta dy + \gamma dz$ will be a perfect differential; so that our hypothesis accounts for the distribution of the lines of force.

It is none too easy to get this part of the picture clear. How the resistance occurs is not explained. Resistance is supposed to be met when the electrical particles pass from one "molecule of the medium" to the next, but within one molecule the resistance is zero. The difference between conductors and insulators is that in the former the particles can pass from one molecule to another but in the latter, including presumably space, they are confined to one molecule. When a current is maintained in a conductor the flow of electricity sets the cells of the medium revolving, rather like the action of a rack on a toothed pinion which meshes with it. The relation of the rotation of the vortices to the current producing it is illustrated in Fig. 11 which is copied from the paper.

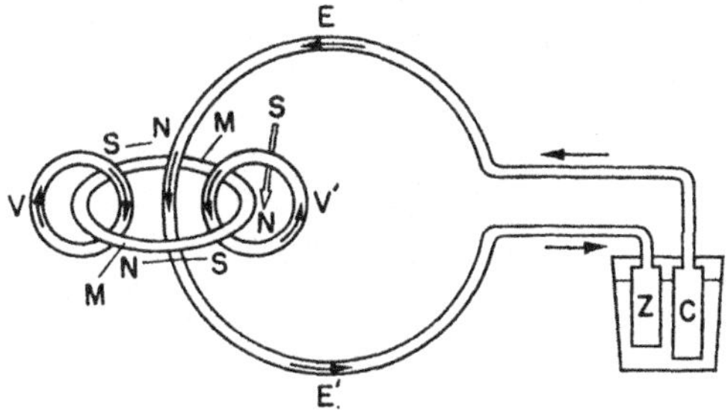


FIG. 11.

The next task is to account for Faraday's laws for the induction of currents and this is done by calculating the energy of the medium. The calculation is made by considering the displacement of two systems of permanent magnetism relative to each other and, though it is couched in much more general terms, it is, in essence, identical to that we have already given in Chapter 4 and the result is the same. The energy $E$ per unit volume is

$$E = \frac{\mu v^2}{8\pi} = \frac{\mu}{8\pi}(\alpha^2 + \beta^2 + \gamma^2). \tag{3}$$

This, as we have already seen, is sufficient to lead to the law of the induction of currents.

Maxwell defines a vector potential *F*, *G*, *H*, so that

$$\mu\alpha = \text{curl}\, \boldsymbol{F} \quad \text{and} \quad \text{div}\, \boldsymbol{F} = 0$$

which gives, in the usual way, Faraday's law for the electromotive force *P*, *Q*, *R*, produced by induction as

$$\boldsymbol{P} = \frac{d\boldsymbol{F}}{dt}.$$

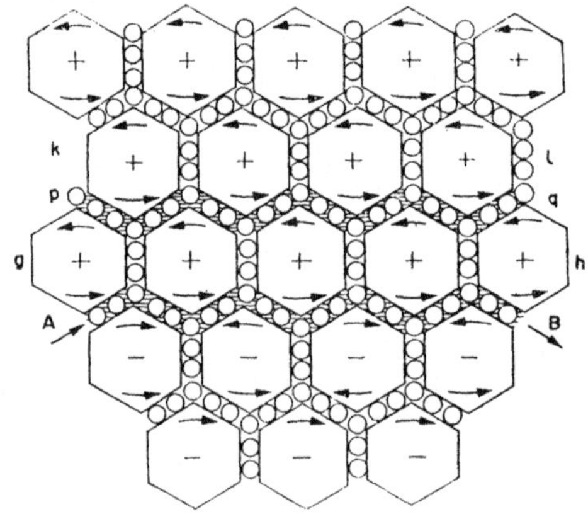


FIG. 12.

A qualitative picture of what is supposed to happen is built up on the basis of Fig. 12, also taken from the paper. Suppose a current to be started from *A* to *B*. The vortices in the row *gh* will be set rotating and these will set the particles in the line *pq* moving in the opposite direction to the current in *AB*. If the induced current is impeded by a resistance the vortices in the row *kl* will be set rotating and when their velocity has become equal to that of the row *gh*, the current in *pq* will cease. When the primary current from *A* to *B* is stopped, the vortices in the row *gh* will stop, so that the continued rotation of the vortices in the row *kl* will lead to an induced current from *p* to *q*, until the energy is exhausted by the work done against the resistance.

A detailed calculation of the effect of deforming a medium containing vortices is then made and the electromotive forces experienced by a body moving with velocity $\boldsymbol{v}$ through a magnetic field shown to be

$$\boldsymbol{P} = \mu\boldsymbol{\alpha} \times \boldsymbol{v} + \frac{d\boldsymbol{F}}{dt} - \text{grad}\, \psi$$

$\psi$ being a scalar.

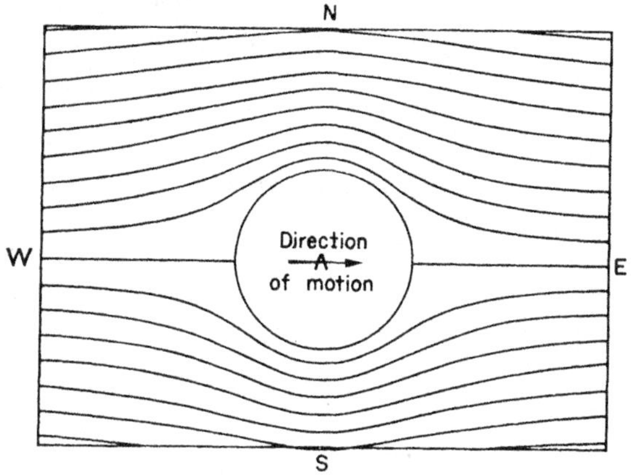


FIG. 13.

What is imagined to happen can be seen from Fig. 13. Suppose $A$ is the horizontal section of a vertical wire moving from west to east across a magnetic field running from south to north. In the figure the lines drawn running from east to west are the lines of flow of the medium past the wire, the whole system being given a velocity towards the west to bring the wire to rest. The vortices, the axes of which run north–south, will to the east of the wire be lengthened longitudinally and contracted laterally as they approach the wire. Their rotation will therefore be speeded up. To the west they will be slowed down again. The vortices are assumed to rotate clockwise when looked at from south to north and they will, therefore, be coming upwards on their western sides and going downwards on their eastern. The speeding up of the vortices on the eastern side of the wire will thus urge the electricity in the wire upwards. The vortices to the west of the wire present their downwards motion to it and when they are slowed down the electricity in the wire is again urged upwards.

Particular interest attaches to Part III of this contribution to the theory, which was published in the *Philosophical Magazine* for January–February 1862, and which was given the title "The Theory of Molecular Vortices Applied to Statical Electricity". It contains the first discussion of the displacement current and the first attempt at an electromagnetic theory of light. The aether filled all space and it contained the stuff of electricity. Electrostatic charges arose because of an elastic displacement of the electricity and to produce it required an electromotive force. The displacement was proportional to the force and disappeared when the force was removed.

If $R$ is the electric restoring force exerted by the medium and $h$ the displacement (i.e. the charge carried across unit area at right angles to the movement) Maxwell wrote

$$R = -4\pi E^2 h$$

$E$ being a coefficient depending on the nature of the dielectric. Variation of the displacement, since it took charge from one extremity of the medium to the other, constituted a current the density, $r$, of which was given by

$$r = \frac{dh}{dt}.$$

Consider a slab of uniformly polarized dielectric, as in a parallel plate condenser, with a surface density of charge $h$; the force per unit area on the surface would be

$$-Rh = 4\pi E^2 h^2.$$

If the charge had been measured in electrostatic units and was of magnitude $h'$ the force would have been,

$$4\pi h'^2.$$

Thus, $E = h'/h$, the ratio of the measure of a charge in electrostatic units to its measure in electromagnetic.

Maxwell then calculated the rigidity which the medium would have to possess in order that the observed relation between $R$ and $h$ should be obtained. For this purpose he chose, for some reason which is not apparent, to consider a medium divided into spherical cells, in spite of the fact that this would not allow them to remain in contact with his single layer of electrical ball bearings. His calculation also became more elaborate than his model justified. It is obviously not worth more than an order of magnitude estimate and this can be more easily based upon the much cruder picture of cubical cells. As it leads to the same result as the more complicated version, we will confine our attention to the simpler calculation.

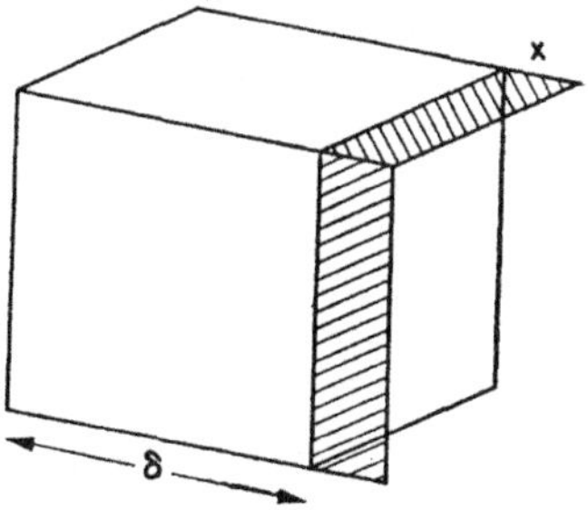


FIG. 14.

First let us estimate the actual distance moved by the electricity in terms of the dimensions of the cells. If $\rho$ is, as before, the charge density on the walls separating the cells and $x$ is the distance through which the electricity moves, we shall have

$$h = \frac{2x\rho}{\delta}$$

where $\delta$ is the side of one of the cells (each cell will contribute a charge of $2x\delta\rho$ as indicated in Fig. 14 over an area of $\delta^2$).

Thus

$$x = \frac{R\delta}{8\pi E^2\rho}.$$

To calculate the coefficient of rigidity $\eta$ consider the medium sheared by the displacement of layers parallel to the plane $z = 0$ as in Fig. 15.

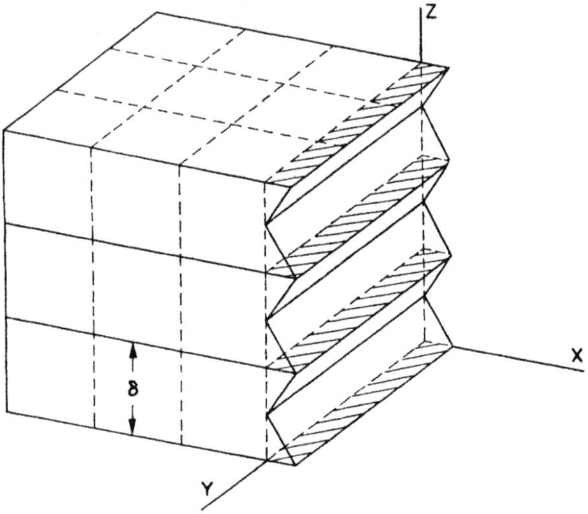


FIG. 15.

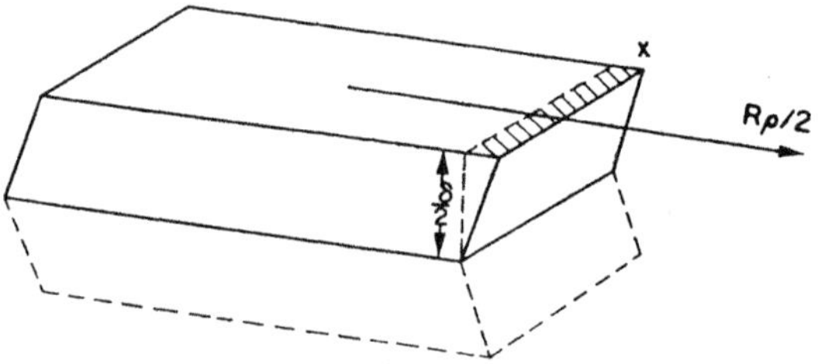


FIG. 16.

We have

$$\eta = \frac{R\rho}{4}\frac{\delta}{x}$$

$$= 2\pi E^2\rho^2.$$

Now we have already found (equation (2)) that

$$\rho = \frac{1}{2\pi}.$$

So that

$$\eta = \frac{E^2}{2\pi}.$$

To calculate the velocity of propagation of waves through the medium we must also know its density, *d*. Maxwell obtained this from the energy of the medium.

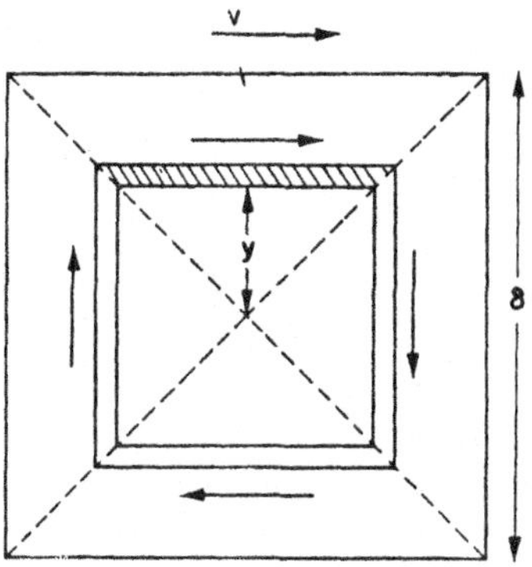


FIG. 17.

Let us assume that the aether circulates in the cubical vortices with a velocity proportional to the distance of its path from the centre as in Fig. 17. The kinetic energy per unit volume will be

$$4\frac{d}{\delta^2}\int_0^{\delta/2} y\,dy\left(\frac{2yv}{\delta}\right)^2 = \frac{v^2 d}{4}$$

where $v$ is the peripheral velocity.

But from equation (3) the energy per unit volume of the medium has been found to be

$$\frac{\mu v^2}{8\pi}$$

giving

$$d = \frac{\mu}{2\pi}$$

We have, therefore, for the velocity of propagation of transverse vibrations through the medium,

$$V = \sqrt{\frac{\eta}{d}} = \frac{E}{\sqrt{\mu}}.$$

This, in a nutshell, is Maxwell's first electromagnetic theory of light. By it he sought to show that it was not necessary to postulate different media for the propagation of light and to account for the phenomena of electrodynamics. The same aether, vortices in which constituted the lines of magnetic force, also would be capable of transmitting transverse elastic disturbances with the velocity of light.

In the fourth and last section of this paper Maxwell gave a theory of the rotation of the plane of polarization of light travelling along the lines of force in a magnetic field. The beam of plane polarized light was resolved into two circularly polarized components rotating in opposite directions. The aetherial vortices speeded up the rotation of one component and slowed down that of the other. The effect was a rotation of the plane of polarization of the resultant, in accordance with observation. Though at the time the theory provided evidence supporting the idea of aetherial vortices, we need hardly concern ourselves further with it here.

One important effect of this investigation of the rotation of the plane of polarization of light, however, was to dispose of Faraday's classification of a vacuum as coming between paramagnetic and diamagnetic substances, so that the behaviour of diamagnetic material could be looked upon as the same as that of a paramagnetic substance surrounded by a medium of greater permeability. M. Verdet discovered that paramagnetic substances have an effect on light opposite to that of diamagnetic substances, and it follows from this, together with Maxwell's theory, that the "molecular" rotations must be in the opposite direction in the two classes of substances. "We must admit", concluded Maxwell, "the diamagnetic state to be the *opposite* of the paramagnetic."

We come, therefore, to Maxwell's third contribution to electromagnetic theory, in which he gave the complete picture more or less as we know it today. As the main part of this paper is reproduced in this volume, brief comments on it only are necessary. The object of this research was to rid the theory of those makeshift and detailed assumptions, like that of the electrical ball bearings, which had previously helped it forward.

> I have on a former occasion [he explained] attempted to describe a particular kind of motion and a particular kind of strain, so arranged as to account for the phenomena. In the present paper I avoid any hypothesis of this kind; and in using such words as electric momentum and electric elasticity. . . . I wish merely to direct the mind of the reader to mechanical phenomena which will assist him in understanding the electrical ones. All such phrases in the present paper are to be considered as illustrative not explanatory.
>
> In speaking of the energy of the field, however, I wish to be understood literally. All energy is the same as mechanical energy. . . .
>
> The only question is Where does it reside? . . .

The Cheshire Cat has started to disappear. Its detail has gone but the substance of it still remains. There seems to be little doubt that Maxwell continued to approach the whole problem from Faraday's point of view of a medium, and, in particular, arrived at the idea of the displacement current in this way, though, as we shall see in a moment, he was well aware that the displacement current was necessitated in order that his equations for the magnetic field might be consistent—the reason most frequently put forward to support the concept today, now that the medium itself has completed its disappearance.

The order in which the subject is developed in the third paper is to accept the laws of electromagnetic induction as provided from experiment by Faraday and from them to deduce those of the mechanical action of currents. Granting Maxwell's point of view, this order is to a large extent arbitrary. Both laws are first obtained for two circuits in terms of three arbitrary constants, $L, M, N$, which can then be identified either from the laws of induction, as Maxwell suggests in this paper, in which case the law of the mechanical action of currents can be presented as a deduction, or the process can be inverted and the constants identified from the laws of mechanical action and those of the induction of currents obtained deductively. It is possible that it is a little easier to identify the constants $L$ and $N$ in Maxwell's way, so that his order is perhaps to be preferred, but neither process is altogether free from assumption—the situation was discussed in Chapter IV.

This paper contains the first complete statement of the equations of the electromagnetic field, with which Maxwell's name has since been associated. He gives, in fact, eight of them and they are all expressed in electromagnetic units.

Equations A of the paper, which in vector notation, taking Maxwell's first symbol to represent the whole vector, were

$$\boldsymbol{p}' = \boldsymbol{p} + \frac{d\boldsymbol{f}}{dt}.$$

This equation expresses the total current $\boldsymbol{p}'$ as the sum of the conduction current $\boldsymbol{p}$ and the displacement current $d\boldsymbol{f}/dt$. It was arrived at, as we have seen, as a result of the mathematical expression of Faraday's concepts concerning the function of the medium. Today with the disappearance of the medium it is usually approached differently. Ampère's experiments for closed circuits lead to the work law, which is expressed in the usual notation by the equation

$$\text{curl}\,\boldsymbol{H} = \frac{4\pi}{c}\boldsymbol{j}$$

This entails that

$$\text{div}\,\boldsymbol{j} = 0$$

which is merely a statement of the fact that the law applies only to closed circuits, in which electricity does not accumulate at any point. In the case of an open circuit, as in that of the charging of a condenser, for example,

$$\text{div}\,\boldsymbol{j} = -\frac{d\rho}{dt}$$

which is not zero, but the rate at which charge flows away. At this point it is sometimes stated that the two equations are inconsistent and therefore require modification. This, of course, is true as the equations are written, but they are only inconsistent in the fact that the same symbol, $j$, has been used for two different things—for a steady circuital current in the first case and for a limited current in the second. The difficulty is another example of that which has been encountered before in writing Ampère's law of action of current elements (of which curl $H = 4\pi j/c$ is one expression) in differential form. So long as attention is concentrated upon elements it is impossible to distinguish between circuital and finite currents. Maxwell's way out of the difficulty was to make all currents circuital. Since

$$4\pi \frac{d\rho}{dt} = \operatorname{div} \frac{dD}{dt}$$

$$\operatorname{div}\left(j + \frac{1}{4\pi}\frac{dD}{dt}\right) = 0$$

and, if it were postulated that the term $\frac{1}{4\pi}\frac{dD}{dt}$ represents a current possessing the same properties as any other current, then

$$\operatorname{curl} \boldsymbol{H} = \frac{1}{c}\frac{dD}{dt} + \frac{4\pi}{c}\boldsymbol{j}$$

and the equation might not have to be restricted to cases of closed circuits only. The argument is in no sense a proof. The introduction of the displacement current remains a postulate and it is rather doubtful if the argument even renders it more plausible. The first law of circulation was arrived at through the consideration of experiments made entirely upon closed circuits and there is no *a priori* reason why it should apply to cases where the circuits are not closed. The only way in which the postulate can be justified is through an examination of its consequences, that is to say, in effect through a study of radiation.

Maxwell was aware of this argument. In Article 607 of the *Treatise* he makes this remark:

> We have very little experimental evidence relating to the direct electromagnetic action of currents due to the variation of electric displacement in dielectrics, but the extreme difficulty of reconciling the laws of electromagnetism with the existence of electric currents which are not closed is

one reason among many why we must admit the existence of transient currents due to the variation of the displacement.

Another way out of the difficulty would be to employ Heaviside circuital current elements in place of the line elements of Ampère. These, like Maxwell's hypothesis of a displacement current, include a postulated completion of the circuit through the surrounding space and reduce to the same thing in the end.

Equations B are, in vector form,

$$\mu\boldsymbol{\alpha} = \text{curl } \boldsymbol{F}$$

which relates the magnetic induction $\mu\boldsymbol{\alpha}$ to the vector potential, $\boldsymbol{F}$. Taking the divergence of both sides this is equivalent to the modern equation,

$$\text{div } \boldsymbol{B} = 0. \tag{4}$$

Maxwell's equations C are

$$\text{curl } \boldsymbol{\alpha} = 4\pi \boldsymbol{p}'$$

which is the work law for the total current and is equivalent to

$$\text{curl } \boldsymbol{H} = \frac{4\pi}{c} j + \frac{1}{c} \frac{dD}{dt}. \tag{5}$$

The equations D are for the electromotive force at a point in a circuit moving with a velocity $v$ and are

$$\boldsymbol{P} = \boldsymbol{\alpha} \times \boldsymbol{v} - \frac{d\boldsymbol{F}}{dt} - \text{grad } \psi.$$

If we assume the circuit at rest and take the curl of both sides we have, in the modern form,

$$\text{curl } \boldsymbol{E} = -\frac{1}{c} \frac{d\boldsymbol{B}}{dt}. \tag{6}$$

Equations E and F are simply the equations for the displacement and current in terms of Maxwell's "electromotive force". Equation G is

$$e + \text{div} \boldsymbol{f} = 0.$$

Here $e$ is the free electricity in unit volume and is the charge carried into an element per unit volume by the displacement, $f$. If we put a charge $\rho$ per unit volume into an element, then the displacement carries the same charge away so that we have the usual equation,

$$\operatorname{div} \boldsymbol{D} = 4\pi\rho \tag{7}$$

if $\rho$ represents, as usual, the charge residing in a body. Maxwell illustrated this as follows:†

> If a charge $e$ is uniformly distributed over the surface of a sphere, the resultant intensity at any point of the medium distant $r$ from the centre . . . is proportional to $e/r^2$. The displacement in the medium will therefore be proportional to $e/r^2$, so that if we draw a sphere of radius $r$ concentric with the first its surface will vary as $r^2$ and the total displacement through it will be proportional to $e$ and independent of the radius. If $V_1$ and $V_2$ are the potentials of the inner and outer spheres respectively the work done in increasing the displacement by $dE$ will be $(V_1 - V_2)\,dE$. If we now take the outer sphere to infinity $V_1$ becomes the potential of the charged sphere and $V_2$ zero. The work is thus $VdE$. But the work is also $Vde$ where $de$ is the increase in the charge of the sphere. Thus if we take the view that electrical energy resides in the medium $dE = de$. The displacement outwards through any spherical surface concentric with the sphere is equal to the charge on the sphere. It follows from this that the displacement current renders any otherwise finite current continuous and equivalent to a current in a closed ciruit.

It only remains to say a very few words about subsequent history. The classical aether has disappeared and with it the physical picture from which Faraday and Maxwell derived their theory. At the time of its going a common feeling undoubtedly was of something akin to consternation at the loss of a fundamental understanding and explanation of electrical phenomena. All that was left was the grin without the cat. But what has in fact been lost? Speaking of mechanical explanations in physical theories, Henri Poincaré wrote in his *Science and Hypothesis*:‡

> If the principle of least action cannot be satisfied, no mechanical explanation is possible; if it can be satisfied, there is not only one explanation, but an unlimited number. . . . Maxwell recognised that electrical phenomena satisfy the principle of least action. He was then certain of a mechanical explanation. If he had expounded this theory at the beginning

† *Treatise*, Article 60.
‡ p. 221, Dover edition.

> of his first volume, instead of relegating it to a corner in the second, it would not have escaped the attention of most readers. If, therefore, a phenomenon allows of a complete mechanical explanation, it allows of an unlimited number of others, which will equally take into account all the particulars revealed by experiment. ...
>
> How shall we choose from all the possible explanations one, in which the help of experiment will be wanting? The day will perhaps come when physicists will no longer concern themselves with questions which are inaccessible to positive methods, and will leave them to the metaphysicians. That day has not yet come; man does not so easily resign himself to remaining for ever ignorant of the causes of things.

Science, however, can never provide ultimate explanations but only describe. No problem can be pushed more than one stage further back and the question of how the explanation is itself to be explained will always remain. What the aether theory did, as Maxwell emphasizes over and over again, was to provide an analogy in terms of which electromagnetic phenomena could be grasped. It helped enormously in thinking about electricity to emphasize the similarities that existed between it and fluids and elastic solids, the behaviour of which was already familiar. But once familiarity was achieved with the phenomena of electricity on their own, the necessity for the props derived from other fields became less insistent. They can still be employed if desired.

In answer to the question "What is Maxwell's theory?" Hertz gave the answer: "Maxwell's theory is Maxwell's system of equations. Every theory which leads to the same system of equations and therefore comprises the same possible phenomena, I would consider as being a form or special case of Maxwell's theory." He went on to say, however: "If we wish to lend more colour to the theory, there is nothing to prevent us from supplementing all this and aiding our powers of imagination by concrete representations of various conceptions as to the nature of electric polarisation, the electric current, etc." Electricity still behaves, in many ways *as if* it were a fluid dispersed in an elastic medium throughout space, but the position has lost its compulsion. We are content to deal with the electric field directly on its own and in fact have employed it as the familiar side in other analogies to help in the exploration of other less well understood phenomena.

The equivalence of a varying electric field to a current at each point of space, which is the essence of Maxwell's theory, can be treated as a postulate to be justified by its consequences, without attaching to it any physical picture to account for its occurrence. As a first approach to Maxwell's theory, however, there is little doubt that to treat the theory as a piece of pure mathematics, with a set of far from self-obvious axioms provided gratuitously at the outset, with no discussion of their grounding, is unsatisfactory to many. It furnishes no insight into the processes of thought which gave rise to the theory. It is not necessary to introduce any particular mechanical picture into the *logical* structure of the completed theory but a historical study of the way in which the theory developed can be more satisfying than a mere acceptance of it as a *fait accompli.*

We have now traced in outline in this book the development of electrical theory from the erection of the scaffolding by Faraday to the completion of the structure by Maxwell. To quote Poincaré on Maxwell once again in conclusion:

> He throws into relief the essential—i.e. what is common to all theories; everything that suits only a particular theory is passed over almost in silence. The reader therefore finds himself in the presence of form nearly devoid of matter, which at first he is tempted to take as a fugitive and unassailable phantom. But the efforts he is thus compelled to make force him to think, and eventually he sees that there is often something rather artificial in the theoretical "aggregates" which he once admired.

# PART 2

# *Extracts*

*1. Michael Faraday*
*Experimental Researches in Electricity*

# First Series

§ 1. *On the Induction of Electric Currents.* § 2. *On the Evolution of Electricity from Magnetism.* § 3. *On a new Electrical Condition of Matter.* § 4. *On* Arago's *Magnetic Phenomena.*

[Read November 24, 1831.]

1. The power which electricity of tension possesses of causing an opposite electrical state in its vicinity has been expressed by the general term Induction; which, as it has been received into scientific language, may also, with propriety, be used in the same general sense to express the power which electrical currents may possess of inducing any particular state upon matter in their immediate neighbourhood, otherwise indifferent. It is with this meaning that I purpose using it in the present paper.

2. Certain effects of the induction of electrical currents have already been recognised and described: as those of magnetization; Ampère's experiments of bringing a copper disc near to a flat spiral; his repetition with electro-magnets of Arago's extraordinary experiments, and perhaps a few others. Still it appeared unlikely that these could be all the effects which induction by currents could produce; especially as, upon dispensing with iron, almost the whole of them disappear, whilst yet an infinity of bodies, exhibiting definite phenomena of induction with electricity of tension, still remain to be acted upon by the induction of electricity in motion.

3. Further: Whether Ampère's beautiful theory were adopted, or any other, or whatever reservation were mentally made, still it appeared very extraordinary, that as every electric current was

accompanied by a corresponding intensity of magnetic action at right angles to the current, good conductors of electricity, when placed within the sphere of this action, should not have any current induced through them, or some sensible effect produced equivalent in force to such a current.

4. These considerations, with their consequence, the hope of obtaining electricity from ordinary magnetism, have stimulated me at various times to investigate experimentally the inductive effect of electric currents. I lately arrived at positive results; and not only had my hopes fulfilled, but obtained a key which appeared to me to open out a full explanation of Arago's magnetic phenomena, and also to discover a new state, which may probably have great influence in some of the most important effects of electric currents.

5. These results I purpose describing, not as they were obtained, but in such a manner as to give the most concise view of the whole.

## § 1. Induction of Electric Currents

6. About twenty-six feet of copper wire one twentieth of an inch in diameter were wound round a cylinder of wood as a helix, the different spires of which were prevented from touching by a thin interposed twine. This helix was covered with calico, and then a second wire applied in the same manner. In this way twelve helices were superposed, each containing an average length of wire of twenty-seven feet, and all in the same direction. The first, third, fifth, seventh, ninth, and eleventh of these helices were connected at their extremities end to end, so as to form one helix; the others were connected in a similar manner; and thus two principal helices were produced, closely interposed, having the same direction, not touching anywhere, and each containing one hundred and fifty-five feet in length of wire.

7. One of these helices was connected with a galvanometer, the other with a voltaic battery of ten pairs of plates four inches square, with double coppers and well charged; yet not the slightest sensible reflection of the galvanometer-needle could be observed.

8. A similar compound helix, consisting of six lengths of copper and six of soft iron wire, was constructed. The resulting iron helix contained two hundred and fourteen feet of wire, the resulting copper helix two hundred and eight feet; but whether the current from the trough was passed through the copper or the iron helix, no effect upon the other could be perceived at the galvanometer.

9. In these and many similar experiments no difference in action of any kind appeared between iron and other metals.

10. Two hundred and three feet of copper wire in one length were coiled round a large block of wood; other two hundred and three feet of similar wire were interposed as a spiral between the turns of the first coil, and metallic contact everywhere prevented by twine. One of these helices was connected with a galvanometer, and the other with a battery of one hundred pairs of plates four inches square, with double coppers, and well charged. When the contact was made, there was a sudden and very slight effect at the galvanometer, and there was also a similar slight effect when the contact with the battery was broken. But whilst the voltaic current was continuing to pass through the one helix, no galvanometrical appearances nor any effect like induction upon the other helix could be perceived, although the active power of the battery was proved to be great, by its heating the whole of its own helix, and by the brilliancy of the discharge when made through charcoal.

11. Repetition of the experiments with a battery of one hundred and twenty pairs of plates produced no other effects; but it was ascertained, both at this and the former time, that the slight deflection of the needle occurring at the moment of completing the connexion, was always in one direction, and that the equally slight deflection produced when the contact was broken, was in the other direction; and also, that these effects occurred when the first helices were used.

12. The results which I had by this time obtained with magnets led me to believe that the battery current through one wire, did, in reality, induce a similar current through the other wire, but that it continued for an instant only, and partook more of the nature of the electrical wave passed through from the shock of a

common Leyden jar than of the current from a voltaic battery, and therefore might magnetise a steel needle, although it scarcely affected the galvanometer.

13. This expectation was confirmed; for on substituting a small hollow helix, formed round a glass tube, for the galvanometer, introducing a steel needle, making contact as before between the battery and the inducing wire, and then removing the needle before the battery contact was broken, it was found magnetised.

14. When the battery contact was first made, then an unmagnetised needle introduced into the small indicating helix (13.), and lastly the battery contact broken, the needle was found magnetised to an equal degree apparently as before; but the poles were of the contrary kind.

15. The same effects took place on using the large compound helices first described.

16. When the unmagnetised needle was put into the indicating helix, before contact of the inducing wire with the battery, and remained there until the contact was broken, it exhibited little or no magnetism; the first effect having been nearly neutralised by the second. The force of the induced current upon making contact was found always to exceed that of the induced current at breaking of contact; and if therefore the contact was made and broken many times in succession, whilst the needle remained in the indicating helix, it at last came out not unmagnetised, but a needle magnetised as if the induced current upon making contact had acted alone on it. This effect may be due to the accumulation (as it is called) at the poles of the unconnected pile, rendering the current upon first making contact more powerful than what it is afterwards, at the moment of breaking contact.

17. If the circuit between the helix or wire under induction and the galvanometer or indicating spiral was not rendered complete *before* the connexion between the battery and the inducing wire was completed or broken, then no effects were perceived at the galvanometer. Thus, if the battery communications were first made, and then the wire under induction connected

with the indicating helix, no magnetising power was there exhibited. But still retaining the latter communications, when those with the battery were broken, a magnet was formed in the helix, but of the second kind, i.e. with poles indicating a current in the same direction to that belonging to the battery current, or to that always induced by that current at its cessation.

18. In the preceding experiments the wires were placed near to each other, and the contact of the inducing one with the battery made when the inductive effect was required; but as the particular action might be supposed to be exerted only at the moments of making and breaking contact, the induction was produced in another way. Several feet of copper wire were stretched in wide zigzag forms, representing the letter W, on one surface of a broad board; a second wire was stretched in precisely similar forms on a second board, so that when brought near the first, the wires should everywhere touch, except that a sheet of thick paper was interposed. One of these wires was connected with the galvanometer, and the other with a voltaic battery. The first wire was then moved towards the second, and as it approached, the needle was deflected. Being then removed, the needle was deflected in the opposite direction. By first making the wires approach and then recede, simultaneously with the vibrations of the needle, the latter soon became very extensive; but when the wires ceased to move from or towards each other, the galvanometer-needle soon came to its usual position.

19. As the wires approximated, the induced current was in the *contrary* direction to the inducing current. As the wires receded, the induced current was in the *same* direction as the inducing current. When the wires remained stationary, there was no induced current.

20. When a small voltaic arrangement was introduced into the circuit between the galvanometer and its helix or wire, so as to cause a permanent deflection of 30° or 40°, and then the battery of one hundred pairs of plates connected with the inducing wire, there was an instantaneous action as before; but the galvanometer-needle immediately resumed and retained its place unaltered, notwithstanding the continued contact of the inducing wire with

the trough: such was the case in whichever way the contacts were made.

21. Hence it would appear that collateral currents, either in the same or in opposite directions, exert no permanent inducing power on each other, affecting their quantity or tension.

22. I could obtain no evidence by the tongue, by spark, or by heating fine wire or charcoal, of the electricity passing through the wire under induction; neither could I obtain any chemical effects, though the contacts with metallic and other solutions were made and broken alternately with those of the battery, so that the second effect of induction should not oppose or neutralise the first.

23. This deficiency of effect is not because the induced current of electricity cannot pass fluids, but probably because of its brief duration and feeble intensity; for on introducing two large copper plates into the circuit on the induced side, the plates being immersed in brine, but prevented from touching each other by an interposed cloth, the effect at the indicating galvanometer, or helix, occurred as before. The induced electricity could also pass through a voltaic trough (20.). When, however, the quantity of interposed fluid was reduced to a drop, the galvanometer gave no indication.

24. Attempts to obtain similar effects by the use of wires conveying ordinary electricity were doubtful in the results. A compound helix similar to that already described, containing eight elementary helices, was used. Four of the helices had their similar ends bound together by wire, and the two general terminations thus produced connected with the small magnetising helix containing an unmagnetised needle. The other four helices were similarly arranged, but their ends connected with a Leyden jar. On passing the discharge, the needle was found to be a magnet; but it appeared probable that a part of the electricity of the jar had passed off to the small helix, and so magnetised the needle. There was indeed no reason to expect that the electricity of a jar possessing as it does great tension, would not diffuse itself through all the metallic matter interposed between the coatings.

25. Still it does not follow that the discharge of ordinary electricity through a wire does not produce analogous phenomena to those arising from voltaic electricity; but as it appears impossible to separate the effects produced at the moment when the discharge begins to pass, from the equal and contrary effects produced when it ceases to pass, inasmuch as with ordinary electricity these periods are simultaneous, so there can be scarcely any hope that in this form of the experiment they can be perceived.

26. Hence it is evident that currents of voltaic electricity present phenomena of induction somewhat analogous to those produced by electricity of tension, although, as will be seen hereafter, many differences exist between them. The result is the production of other currents, (but which are only momentary,) parallel, or tending to parallelism, with the inducing current. By reference to the poles of the needle formed in the indicating helix and to the deflections of the galvanometer-needle, it was found in all cases that the induced current, produced by the first action of the inducing current, was in the contrary direction to the latter, but that the current produced by the cessation of the inducing current was in the same direction. For the purpose of avoiding periphrasis, I propose to call this action of the current from the voltaic battery, *volta-electric induction.* The properties of the second wire, after induction has developed the first current, and whilst the electricity from the battery continues to flow through its inducing neighbour, constitute a peculiar electric condition, the consideration of which will be resumed hereafter (60.). All these results have been obtained with a voltaic apparatus consisting of a single pair of plates.

## § 2. Evolution of Electricity from Magnetism

27. A welded ring was made of soft round bar-iron, the metal being seven-eighths of an inch in thickness, and the ring six inches in external diameter. Three helices were put round one part of

this ring, each containing about twenty-four feet of copper wire one twentieth of an inch thick; they were insulated from the iron and each other, and superposed in the manner before described, occupying about nine inches in length upon the ring. They could be used separately or conjointly; the group may be distinguished by the letter A (fig. 1.). On the other part of the ring about sixty

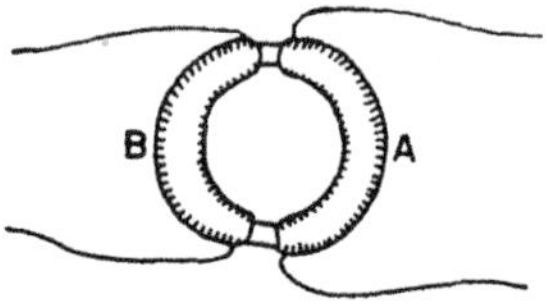


FIG. 1.

feet of similar copper wire in two pieces were applied in the same manner, forming a helix B, which had the same common direction with the helices of A, but being separated from it at each extremity by about half an inch of the uncovered iron.

28. The helix B was connected by copper wires with a galvanometer three feet from the ring. The helices of A were connected end to end so as to form one common helix, the extremities of which were connected with a battery of ten pairs of plates four inches square. The galvanometer was immediately affected, and to a degree far beyond what has been described when with a battery of tenfold power helices *without iron* were used; but though the contact was continued, the effect was not permanent, for the needle soon came to rest in its natural position, as if quite indifferent to the attached electro-magnetic arrangement. Upon breaking the contact with the battery, the needle was again powerfully deflected, but in the contrary direction to that induced in the first instance.

29. Upon arranging the apparatus so that B should be out of use, the galvanometer be connected with one of the three wires of A, and the other two made into a helix through which the current from the trough was passed, similar but rather more powerful effects were produced.

30. When the battery contact was made in one direction, the galvanometer-needle was deflected on the one side; if made in the other direction, the deflection was on the other side. The deflection on breaking the battery contact was always the reverse of that produced by completing it. The deflection on making a battery contact always indicated an induced current in the opposite direction to that from the battery; but on breaking the contact the deflection indicated an induced current in the same direction as that of the battery. No making or breaking of the contact at B side, or in any part of the galvanometer circuit, produced any effect at the galvanometer. No continuance of the battery current caused any deflection of the galvanometer-needle. As the above results are common to all these experiments, and to similar ones with ordinary magnets to be hereafter detailed, they need not be again particularly described.

31. Upon using the power of one hundred pairs of plates with this ring, the impulse at the galvanometer, when contact was completed or broken, was so great as to make the needle spin round rapidly four or five times, before the air and terrestrial magnetism could reduce its motion to mere oscillation. . . .

34. Another arrangement was then employed connecting the former experiments on volta-electric induction with the present. A combination of helices like that already described was constructed upon a hollow cylinder of pasteboard: there were eight lengths of copper wire, containing altogether 220 feet; four of these helices were connected end to end, and then with the galvanometer; the other intervening four were also connected end to end, and the battery of one hundred pairs discharged through them. In this form the effect on the galvanometer was hardly sensible, though magnets could be made by the induced current. But when a soft iron cylinder seven eighths of an inch thick, and twelve inches long, was introduced into the pasteboard tube, surrounded by the helices, then the induced current affected the galvanometer powerfully and with all the phenomena just described. It possessed also the power of making magnets with more energy, apparently, than when no iron cylinder was present.

35. When the iron cylinder was replaced by an equal cylinder of copper, no effect beyond that of the helices alone was produced. The iron cylinder arrangement was not so powerful as the ring arrangement already described.

36. Similar effects were then produced by *ordinary magnets*: thus the hollow helix just described had all its elementary helices connected with the galvanometer by two copper wires, each five feet in length; the soft iron cylinder was introduced into its axis; a couple of bar magnets, each twenty-four inches long, were arranged with their opposite poles at one end in contact, so as to resemble a horse-shoe magnet, and then contact made between the other poles and the ends of the iron cylinder, so as to convert it for the time into a magnet (fig. 2): by breaking the magnetic contacts, or reversing them, the magnetism of the iron cylinder could be destroyed or reversed at pleasure.

FIG. 2.

37. Upon making magnetic contact, the needle was deflected; continuing the contact, the needle became indifferent, and resumed its first position; on breaking the contact, it was again deflected, but in the opposite direction to the first effect, and then it again became indifferent. When the magnetic contacts were reversed the deflections were reversed.

38. When the magnetic contact was made, the deflection was such as to indicate an induced current of electricity in the opposite direction to that fitted to form a magnet, having the same polarity as that really produced by contact with the bar magnets. Thus when the marked and unmarked poles were placed as in fig. 3, the current in the helix was in the direction represented, P being supposed to be the end of the wire going to the positive pole of the battery, or that end towards which the zinc plates face, and N the negative wire. Such a current would have converted the cylinder into a magnet of the opposite kind to that formed by

contact with the poles A and B; and such a current moves in the opposite direction to the currents which in M. Ampère's beautiful theory are considered as constituting a magnet in the position figured.

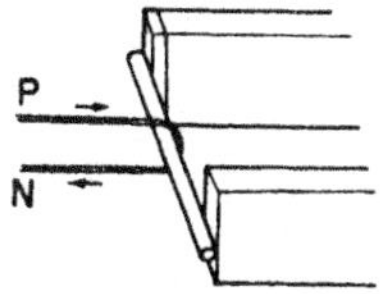


FIG. 3.

39. But as it might be supposed that in all the preceding experiments of this section, it was by some peculiar effect taking place during the formation of the magnet, and not by its mere virtual approximation, that the momentary induced current was excited, the following experiment was made. All the similar ends of the compound hollow helix were bound together by copper wire, forming two general terminations, and these were connected with the galvanometer. The soft iron cylinder was removed, and a cylindrical magnet, three quarters of an inch in diameter and eight inches and a half in length, used instead. One end of this magnet was introduced into the axis of the helix (fig. 4), and then, the

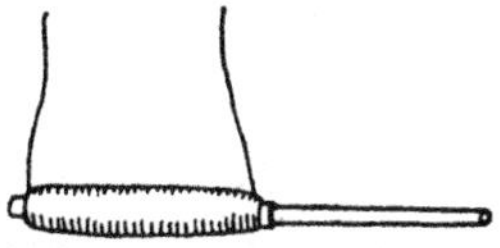

FIG. 4.

galvanometer-needle being stationary, the magnet was suddenly thrust in; immediately the needle was deflected in the same direction as if the magnet had been formed by either of the two preceding processes. Being left in, the needle resumed its first position, and then the magnet being withdrawn the needle was deflected in the opposite direction. These effects were not great; but by introducing and withdrawing the magnet, so that the impulse each time should be added to those previously communicated to the needle, the latter could be made to vibrate through an arc of 180° or more.

40. In this experiment the magnet must not be passed entirely through the helix, for then a second action occurs. When the magnet is introduced, the needle at the galvanometer is deflected in a certain direction; but being in, whether it be pushed quite through or withdrawn, the needle is deflected in a direction the reverse of that previously produced. When the magnet is passed in and through at one continuous motion, the needle moves one way, is then suddenly stopped, and finally moves the other way.

41. If such a hollow helix as that described be laid east and west (or in any other constant position), and a magnet be retained east and west, its marked pole always being one way; then whichever end of the helix the magnet goes in at, and consequently whichever pole of the magnet enters first, still the needle is deflected the same way: on the other hand, whichever direction is followed in withdrawing the magnet, the deflection is constant, but contrary to that due to its entrance.

42. These effects are simple consequences of the *law* hereafter to be described.

43. When the eight elementary helices were made one long helix, the effect was not so great as in the arrangement described. When only one of the eight helices was used, the effect was also much diminished. All care was taken to guard against any direct action of the inducing magnet upon the galvanometer, and it was found that by moving the magnet in the same direction, and to the same degree on the outside of the helix, no effect on the needle was produced.

44. The Royal Society are in possession of a large compound magnet formerly belonging to Dr. Gowin Knight, which, by permission of the President and Council, I was allowed to use in the prosecution of these experiments: it is at present in the charge of Mr. Christie, at his house at Woolwich, where, by Mr. Christie's kindness, I was at liberty to work; and I have to acknowledge my obligations to him for his assistance in all the experiments

and observations made with it. This magnet is composed of about 450 bar magnets, each fifteen inches long, one inch wide, and half an inch thick, arranged in a box so as to present at one of its extremities two external poles (fig. 5). These poles projected

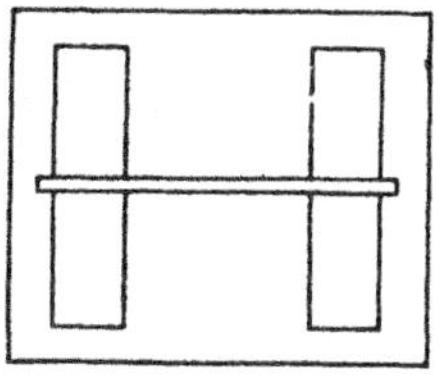

FIG. 5.

horizontally six inches from the box, were each twelve inches high and three inches wide. They were nine inches apart; and when a soft iron cylinder, three quarters of an inch in diameter and twelve inches long, was put across from one to the other, it required a force of nearly one hundred pounds to break the contact. The pole to the left in the figure is the marked pole*.

45. The indicating galvanometer, in all experiments made with this magnet, was about eight feet from it, not directly in front of the poles, but about 16° or 17° on one side. It was found that on making or breaking the connexion of the poles by soft iron, the instrument was slightly affected; but all error of observation arising from this cause was easily and carefully avoided.

46. The electrical effects exhibited by this magnet were very striking. When a soft iron cylinder thirteen inches long was put through the compound hollow helix, with its ends arranged as two general terminations, these connected with the galvanometer, and the iron cylinder brought in contact with the two poles of the magnet (fig. 5.), so powerful a rush of electricity took place that the needle whirled round many times in succession.

* To avoid any confusion as to the poles of the magnet, I shall designate the pole pointing to the north as the marked pole; I may occasionally speak of the north and south ends of the needle, but do not mean thereby north and south poles. That is by many considered the true north pole of a needle which points to the south; but in this country it is often called the south pole.

47. Notwithstanding this great power, if the contact was continued, the needle resumed its natural position, being entirely uninfluenced by the position of the helix. But on breaking the magnetic contact, the needle was whirled round in the opposite direction with a force equal to the former.

48. A piece of copper plate wrapped *once* round the iron cylinder like a socket, but with interposed paper to prevent contact, had its edges connected with the wires of the galvanometer. When the iron was brought in contact with the poles the galvanometer was strongly affected.

49. Dismissing the helices and sockets, the galvanometer wire was passed over, and consequently only half round the iron cylinder (fig. 6.); but even then a strong effect upon the needle was exhibited, when the magnetic contact was made or broken.

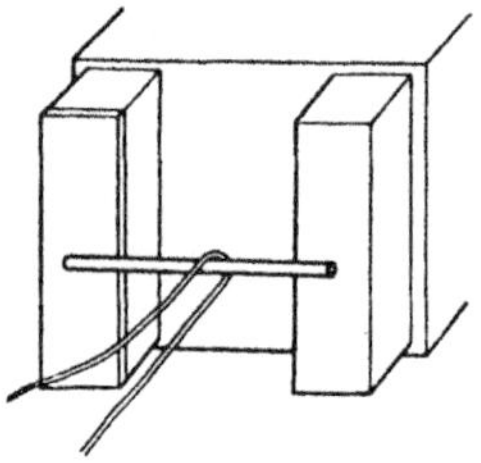

FIG. 6.

50. As the helix with its iron cylinder was brought towards the magnetic poles, but *without making contact*, still powerful effects were produced. When the helix, without the iron cylinder, and consequently containing no metal but copper, was approached to, or placed between the poles, the needle was thrown 80°, 90°, or more, from its natural position. The inductive force was of course greater, the nearer the helix, either with or without its iron cylinder, was brought to the poles; but otherwise the same effects were produced, whether the helix, &c. was or was not brought into contact with the magnet; i. e. no permanent effect on the galvanometer was produced; and the effects of approximation and removal were the reverse of each other.

51. When a bolt of copper corresponding to the iron cylinder was introduced, no greater effect was produced by the helix than without it. But when a thick iron wire was substituted, the magneto-electric induction was rendered sensibly greater.

52. The direction of the electric current produced in all these experiments with the helix, was the same as that already described as obtained with the weaker bar magnets.

53. A spiral containing fourteen feet of copper wire, being connected with the galvanometer, and approximated directly towards the marked pole in the line of its axis, affected the instrument strongly; the current induced in it was in the reverse direction to the current theoretically considered by M. Ampère as existing in the magnet, or as the current in an electro-magnet of similar polarity. As the spiral was withdrawn, the induced current was reversed.

54. A similar spiral had the current of eighty pairs of 4-inch plates sent through it so as to form an electro-magnet, and then the other spiral connected with the galvanometer approximated to it; the needle vibrated, indicating a current in the galvanometer spiral the reverse of that in the battery spiral. On withdrawing the latter spiral, the needle passed in the opposite direction.

55. Single wires, approximated in certain directions towards the magnetic pole, had currents induced in them. On their removal, the currents were inverted. In such experiments the wires should not be removed in directions different to those in which they were approximated; for then occasionally complicated and irregular effects are produced, the causes of which will be very evident in the fourth part of this paper. . . .

57. The various experiments of this section prove, I think, most completely the production of electricity from ordinary magnetism. That its intensity should be very feeble and quantity small, cannot be considered wonderful, when it is remembered that like thermo-electricity it is evolved entirely within the substance of metals retaining all their conducting power. . . .

58. The similarity of action, almost amounting to identity, between common magnets and either electro-magnets or volta-

electric currents, is strikingly in accordance with and confirmatory of M. Ampère's theory, and furnishes powerful reasons for believing that the action is the same in both cases; . . .

## § 3. New Electrical State or Condition of Matter

60. Whilst the wire is subject to either volta-electric or magneto-electric induction, it appears to be in a peculiar state; for it resists the formation of an electrical current in it, whereas, if in its common condition, such a current would be produced; and when left uninfluenced it has the power of originating a current, a power which the wire does not possess under common circumstances. This electrical condition of matter has not hitherto been recognised, but it probably exerts a very important influence in many if not most of the phenomena produced by currents of electricity. For reasons which will immediately appear, I have, after advising with several learned friends, ventured to designate it as the *electro-tonic* state.

61. This peculiar condition shows no known electrical effects whilst it continues; nor have I yet been able to discover any peculiar powers exerted, or properties possessed, by matter whilst retained in this state.

83. Upon obtaining electricity from magnets by the means already described (36. 46.), I hoped to make the experiment of M. Arago a new source of electricity; and did not despair, by reference to terrestrial magneto-electric induction, of being able to construct a new electrical machine. Thus stimulated, numerous experiments were made with the magnet of the Royal Society at Mr. Christie's house, in all of which I had the advantage of his assistance. As many of these were in the course of the investigation superseded by more perfect arrangements, I shall consider myself at liberty to rearrange them in a manner calculated to convey most readily what appears to me to be a correct view of the nature of the phenomena.

84. The magnet has been already described (44.). To concentrate the poles, and bring them nearer to each other, two iron or

steel bars, each about six or seven inches long, one inch wide, and half an inch thick, were put across the poles as in fig. 7, and being supported by twine from slipping, could be placed as near to or far from each other as was required. Occasionally two bars of soft iron were employed, so bent that when applied, one to each pole, the two smaller resulting poles were vertically over each other, either being uppermost at pleasure.

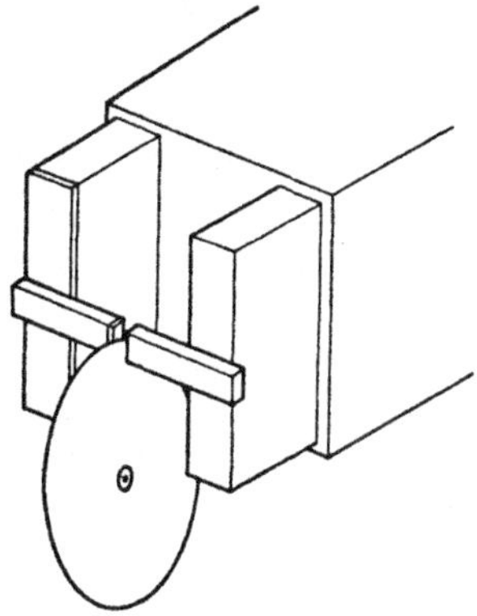

FIG. 7.

85. A disc of copper, twelve inches in diameter, and about one fifth of an inch in thickness, fixed upon a brass axis, was mounted in frames so as to allow of revolution either vertically or horizontally, its edge being at the same time introduced more or less between the magnetic poles (fig. 7.). The edge of the plate was well-amalgamated for the purpose of obtaining a good but moveable contact, and a part round the axis was also prepared in a similar manner.

86. Conductors or electric collectors of copper and lead were constructed so as to come in contact with the edge of the copper disc, or with other forms of plates hereafter to be described. These conductors were about four inches long, one third of an inch wide, and one fifth of an inch thick; one end of each was slightly grooved, to allow of more exact adaptation to the somewhat convex edge of the plates, and then amalgamated. Copper

wires, one sixteenth of an inch in thickness, attached, in the ordinary manner, by convolutions to the other ends of these conductors, passed away to the galvanometer.

87. The galvanometer was roughly made, yet sufficiently delicate in its indications. The wire was of copper covered with silk, and made sixteen or eighteen convolutions. Two sewing-needles were magnetized and fixed on to a stem of dried grass parallel to each other, but in opposite directions, and about half an inch apart; this system was suspended by a fibre of unspun silk, so that the lower needle should be between the convolutions of the multiplier, and the upper above them. The latter was by much the most powerful magnet, and gave terrestrial direction to the whole; fig. 8. represents the direction of the wire and of the needles when

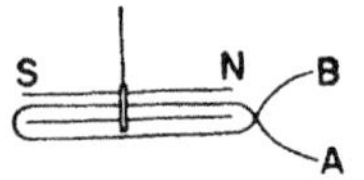


FIG. 8.

the instrument was placed in the magnetic meridian: the ends of the wires are marked A and B for convenient reference hereafter. The letters S and N designate the south and north ends of the needle when affected merely by terrestrial magnetism; the end N is therefore the marked pole. The whole instrument was protected by a glass jar, and stood, as to position and distance relative to the large magnet, under the same circumstances as before.

88. All these arrangements being made, the copper disc was adjusted as in fig. 7, the small magnetic poles being about half an inch apart, and the edge of the plate inserted about half their width between them. One of the galvanometer wires was passed twice or thrice loosely round the brass axis of the plate, and the other attached to a conductor, which itself was retained by the hand in contact with the amalgamated edge of the disc at the part immediately between the magnetic poles. Under these circumstances all was quiescent, and the galvanometer exhibited no effect. But the instant the plate moved, the galvanometer was

influenced, and by revolving the plate quickly the needle could be deflected 90° or more.

89. It was difficult under the circumstances to make the contact between the conductor and the edge of the revolving disc uniformly good and extensive; it was also difficult in the first experiments to obtain a regular velocity of rotation: both these causes tended to retain the needle in a continual state of vibration; but no difficulty existed in ascertaining to which side it was deflected, or generally, about what line it vibrated. Afterwards, when the experiments were made more carefully, a permanent deflection of the needle of nearly 45° could be sustained.

90. Here therefore was demonstrated the production of a permanent current of electricity by ordinary magnets.

91. When the motion of the disc was reversed, every other circumstance remaining the same, the galvanometer needle was deflected with equal power as before; but the deflection was on the opposite side, and the current of electricity evolved, therefore, the reverse of the former.

# Second Series

## § 6. General remarks and illustrations of the Force and Direction of Magneto-electric Induction

193. In the repetition and variation of Arago's experiment by Messrs. Babbage, Herschel, and Harris, these philosophers directed their attention to the differences of force observed amongst the metals and other substances in their action on the magnet. These differences were very great*, and led me to hope that by mechanical combinations of various metals important results might be obtained. The following experiments were therefore made, with a view to obtain, if possible, any such difference of the action of two metals.

194. A piece of soft iron bonnet-wire covered with cotton was laid bare and cleaned at one extremity, and there fastened by metallic contact with the clean end of a copper wire. Both wires were then twisted together like the strands of a rope, for eighteen or twenty inches; and the remaining parts being made to diverge, their extremities were connected with the wires of the galvanometer. The iron wire was about two feet long, the continuation to the galvanometer being copper.

195. The twisted copper and iron (touching each other nowhere but at the extremity) were then passed between the poles of a powerful magnet arranged horse-shoe fashion (fig. 32.); but not

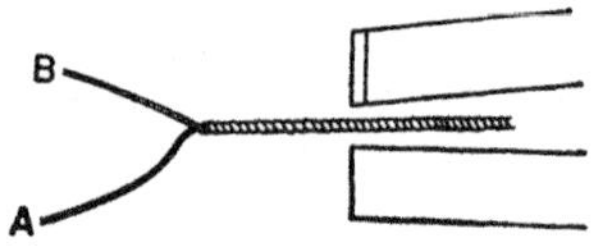


FIG. 32.

* Philosophical Transactions, 1825, p. 472; 1831, p. 78.

the slightest effect was observed at the galvanometer, although the arrangement seemed fitted to show any electrical difference between the two metals relative to the action of the magnet.

204. Experiments were therefore made in which different metals insulated from each other were passed between the poles of the magnet, their opposite ends being connected with the same end of the galvanometer wire, so that the currents formed and led away to the galvanometer should oppose each other; and when considerable lengths of different wires were used, feeble deflections were obtained.

205. To obtain perfectly satisfactory results a new galvanometer was constructed, consisting of two independent coils, each containing eighteen feet of silked copper wire. These coils were exactly alike in shape and number of turns, and were fixed side by side with a small interval between them, in which a double needle could be hung by a fibre of silk exactly as in the former instrument. The coils may be distinguished by the letters KL, and when electrical currents were sent through them in the same direction, acted upon the needle with the sum of their powers; when in opposite directions, with the difference of their powers.

206. The compound helix was now connected, the ends A and B of the iron with A and B ends of galvanometer coil K, and the ends A and B of the copper with B and A ends of galvanometer coil L, so that the currents excited in the two helices should pass in opposite directions through the coils K and L. On introducing a small cylinder magnet within the helices, the galvanometer needle was powerfully deflected. On disuniting the iron helix, the magnet caused with the copper helix alone still stronger deflection in the same direction. On reuniting the iron helix, and unconnecting the copper helix, the magnet caused a moderate deflection in the contrary direction. Thus it was evident that the electric current induced by a magnet in a copper wire was far more powerful than the current induced by the same magnet in an equal iron wire.

207. To prevent any error that might arise from the greater influence, from vicinity or other circumstances, of one coil on the

needle beyond that of the other, the iron and copper terminations were changed relative to the galvanometer coils KL, so that the one which before carried the current from the copper now conveyed that from the iron, and vice versa. But the same striking superiority of the copper was manifested as before. This precaution was taken in the rest of the experiments with other metals to be described.

208. I then had wires of iron, zinc, copper, tin, and lead, drawn to the same diameter (very nearly one twentieth of an inch), and I compared exactly equal lengths, namely sixteen feet, of each in pairs in the following manner: The ends of the copper wire were connected with the ends A and B of galvanometer coil K, and the ends of the zinc wire with the terminations A and B of the galvanometer coil L. The middle part of each wire was then coiled six times round a cylinder of soft iron covered with paper, long enough to connect the poles of Daniell's horse-shoe magnet (fig. 33.),

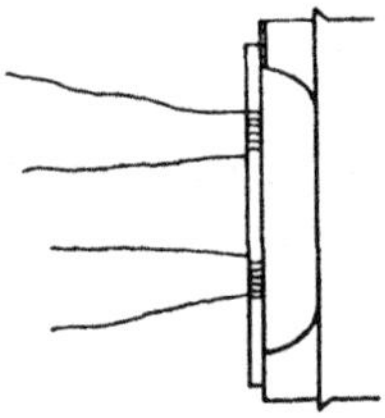

FIG. 33.

so that similar helices of copper and zinc, each of six turns. surrounded the bar at two places equidistant from each other and from the poles of the magnet; but these helices were purposely arranged so as to be in contrary directions, and therefore send contrary currents through the galvanometer coils K and L.

209. On making and breaking contact between the soft iron bar and the poles of the magnet, the galvanometer was strongly affected; on detaching the zinc it was still more strongly affected in the same direction. On taking all the precautions before alluded to, with others, it was abundantly proved that the current induced by the magnet in copper was far more powerful than in zinc.

210. The copper was then compared in a similar manner with tin, lead, and iron, and surpassed them all, even more than it did zinc. The zinc was then compared experimentally with the tin, lead, and iron, and found to produce a more powerful current than any of them. Iron in the same manner proved superior to tin and lead. Tin came next, and lead the last.

211. Thus the order of these metals is copper, zinc, iron, tin, and lead. It is exactly their order with respect to conducting power for electricity, and, with the exception of iron, is the order presented by the magneto-rotation experiments of Messrs. Babbage, Herschel, Harris, &c. The iron has additional power in the latter kind of experiments, because of its ordinary magnetic relations, and its place relative to magneto-electric action of the kind now under investigation cannot be ascertained by such trials. In the manner above described it may be correctly ascertained.

212. It must still be observed that in these experiments the whole effect between different metals is not obtained; for of the thirty-four feet of wire included in each circuit, eighteen feet are copper in both, being the wire of the galvanometer coils; and as the whole circuit is concerned in the resulting force of the current, this circumstance must tend to diminish the difference which would appear between the metals if the circuits were of the same substances throughout. In the present case the difference obtained is probably not more than a half of that which would be given if the whole of each circuit were of one metal.

213. These results tend to prove that the currents produced by magneto-electric induction in bodies is proportional to their conducting power. That they are *exactly* proportional to and altogether dependent upon the conducting power, is, I think, proved by the perfect neutrality displayed when two metals or other substances, as acid, water, &c. &c., are opposed to each other in their action. The feeble current which tends to be produced in the worse conductor, has its transmission favoured in the better conductor, and the stronger current which tends to form in the latter has its intensity diminished by the obstruction of the former; and the forces of generation and obstruction are so

perfectly balanced as to neutralize each other exactly. Now as the obstruction is inversely as the conducting power, the tendency to generate a current must be directly as that power to produce this perfect equilibrium.

*Royal Institution,*
*December* 21, 1831.

# Ninth Series

## § 15. On the influence by induction of an Electric Current on itself:— and on the inductive action of Electric Currents generally

[Received December 18, 1834,—Read January 29, 1835.]

1048. THE following investigations relate to a very remarkable inductive action of electric currents, or of the different parts of the same current, and indicate an immediate connexion between such inductive action and the direct transmission of electricity through conducting bodies, or even that exhibited in the form of a spark.

1049. The inquiry arose out of a fact communicated to me by Mr. Jenkin, which is as follows. If an ordinary wire of short length be used as the medium of communication between the two plates of an electromotor consisting of a single pair of metals, no management will enable the experimenter to obtain an electric shock from this wire; but if the wire which surrounds an electro-magnet be used, a shock is felt each time the contact with the electromotor is broken, provided the ends of the wire be grasped one in each hand.

1050. Another effect is observed at the same time, which has long been known to philosophers, namely, that a bright electric spark occurs at the place of disjunction.

1051. A brief account of these results, with some of a corresponding character which I had observed in using long wires, was published in the Philosophical Magazine for 1834*; and I added to them some observations on their nature. Further investigations led me to perceive the inaccuracy of my first notions, and ended in identifying these effects with the phenomena of

* Vol. v. pp. 349. 444.

induction which I had been fortunate enough to develop in the First Series of these Experimental Researches. Notwithstanding this identity, the extension and the peculiarity of the views respecting electric currents which the results supply, lead me to believe that they will be found worthy of the attention of the Royal Society.

1052. The *electromotor* used consisted of a cylinder of zinc introduced between the two parts of a double cylinder of copper, and preserved from metallic contact in the usual way by corks. The zinc cylinder was eight inches high and four inches in diameter. Both it and the copper cylinder were supplied with stiff wires, surmounted by cups containing mercury; and it was at these cups that the contacts of wires, helices, or electro-magnets, used to complete the circuit, were made or broken. These cups I will call G and E throughout the rest of this paper.

1053. Certain *helices* were constructed, some of which it will be necessary to describe. A pasteboard tube had four copper wires, one twenty-fourth of an inch in thickness, wound round it, each forming a helix in the same direction from end to end: the convolutions of each wire were separated by string, and the superposed helices prevented from touching by intervening calico. The lengths of the wires forming the helices were 48, 49·5, 48, and 45 feet. The first and third wires were united together so as to form one consistent helix of 96 feet in length; and the second and fourth wires were similarly united to form a second helix, closely interwoven with the first, and 94·5 feet in length. These helices may be distinguished by the numbers i and ii. They were carefully examined by a powerful current of electricity and a galvanometer, and found to have no communication with each other.

1054. Another helix was constructed upon a similar pasteboard tube, two lengths of the same copper wire being used, each forty-six feet long. These were united into one consistent helix of ninety-two feet, which therefore was nearly equal in value to either of the former helices, but was not in close inductive association with them. It may be distinguished by the number iii.

1055. A fourth helix was constructed of very thick copper wire, being one-fifth of an inch in diameter; the length of wire used was seventy-nine feet, independent of the straight terminal portions.

1056. The principal *electro-magnet* employed consisted of a cylindrical bar of soft iron twenty-five inches long, and one inch and three quarters in diameter, bent into a ring, so that the ends nearly touched, and surrounded by three coils of thick copper wire, the similar ends of which were fastened together; each of these terminations was soldered to a copper rod, serving as a conducting continuation of the wire. Hence any electric current sent through the rods was divided in the helices surrounding the ring, into three parts, all of which, however, moved in the same direction. The three wires may therefore be considered as representing one wire, of thrice the thickness of the wire really used.

1057. Other electro-magnets could be made at pleasure by introducing a soft iron rod into any of the helices described.

1058. The *galvanometer* which I had occasion to use was rough in its construction, having but one magnetic needle, and not at all delicate in its indications.

1059. The effects to be considered *depend on the conductor* employed to complete the communication between the zinc and copper plates of the electromotor; and I shall have to consider this conductor under four different forms: as the helix of an electro-magnet; as an ordinary helix; as a *long* extended wire, having its course such that the parts can exert little or no mutual influence; and as a *short* wire. In all cases the conductor was of copper.

1060. The peculiar effects are best shown by the *electro-magnet*. When it was used to complete the communication at the electromotor, there was no sensible spark on *making* contact, but on *breaking* contact there was a very large and bright spark, with considerable combustion of the mercury. Then, again, with respect to the shock: if the hands were moistened in salt and water, and good contact between them and the wires retained, no shock could be felt upon *making* contact at the electromotor, but a powerful one on *breaking* contact.

1061. When the *helix* i or iii was used as the connecting conductor, there was also a good spark on breaking contact, but none (sensibly) on making contact. On trying to obtain the shock from these helices, I could not succeed at first. By joining the similar ends of i and ii so as to make the two helices equivalent to one helix, having wire of double thickness, I could just obtain the sensation. Using the helix of thick wire the shock was distinctly obtained. On placing the tongue between two plates of silver connected by wires with the parts which the hands had heretofore touched, there was a powerful shock on *breaking* contact, but none on *making* contact.

1062. The power of producing these phenomena exists therefore in the simple helix, as in the electro-magnet, although by no means in the same high degree.

1063. On putting a bar of soft iron into the helix, it became an electro-magnet, and its power was instantly and greatly raised. On putting a bar of copper into the helix, no change was produced, the action being that of the helix alone. The two helices i and ii, made into one helix of twofold length of wire, produced a greater effect than either i or ii alone.

1064. On descending from the helix to the mere *long wire*, the following effects were obtained. A copper wire, 0·18 of an inch in diameter, and 132 feet in length, was laid out upon the floor of the laboratory, and used as the connecting conductor; it gave no sensible spark on making contact, but produced a bright one on breaking contact, yet not so bright as that from the helix. On endeavouring to obtain the electric shock at the moment contact was broken, I could not succeed so as to make it pass through the hands; but by using two silver plates fastened by small wires to the extremity of the principal wire used, and introducing the tongue between those plates, I succeeded in obtaining powerful shocks upon the tongue and gums, and could easily convulse a flounder, an eel, or a frog. None of these effects could be obtained directly from the electromotor, i.e. when the tongue, frog, or fish was in a similar, and therefore comparative manner, interposed in the course of the communication between the zinc and copper plates, separated everywhere else by the acid used to excite the combination, or by air. The bright spark and the shock, produced only on breaking contact, are therefore effects of the same kind as those produced in a higher degree by the helix, and in a still higher degree by the electro-magnet.

1077. Returning to the phenomena in question, the first thought that arises in the mind is, that the electricity circulates with something like *momentum or inertia* in the wire, and that thus a long wire produces effects at the instant the current is stopped, which a short wire cannot produce. Such an explantion is, however, at once set aside by the fact, that the same length of wire produces the effects in very different degrees, according as it is simply extended, or made into a helix, or forms the circuit of an electro-magnet. The experiments to be adduced will still more strikingly show that the idea of momentum cannot apply.

1078. The bright spark at the electromotor, and the shock in the arms, appeared evidently to be due to *one* current in the long wire, divided into two parts by the double channel afforded through the body and through the electromotor; for that the spark was evolved at the place of disjunction with the electromotor, not by any direct action of the latter, but by a force immediately exerted in the wire of communication, seemed to be without doubt. It followed, therefore, that by using a better conductor in place of the human body, the *whole* of this extra current might be made to pass at that place; and thus be separated from that which the electromotor could produce by its immediate action, and its *direction* be examined apart from any interference of the original and originating current. This was found to be true; for on connecting the ends of the principal wire together by a cross wire two or three feet in length, applied just where the hands had felt the shock, the whole of the extra current passed by the new channel, and then no better spark than one producible by a short wire was obtained on disjunction at the electromotor.

1079. The *current* thus separated was examined by galvanometers and decomposing apparatus introduced into the course of this wire. I will always speak of it as the current in the cross wire or wires, so that no mistake, as to its place or origin, may occur. In the wood-cut, Z and C represent the zinc and copper plates of the electromotor; G and E the cups of mercury where contact is made or broken (1052.); A and B the terminations of D, the long

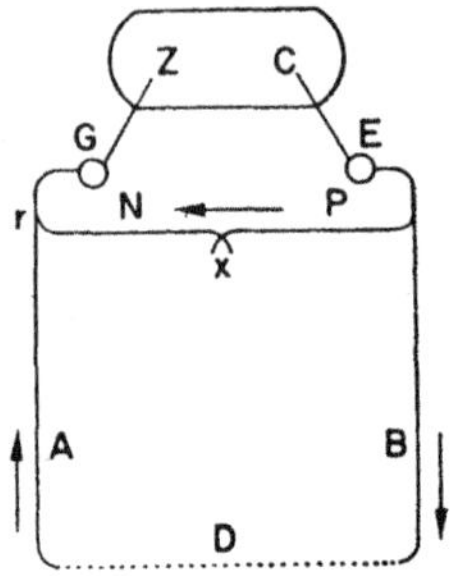


wire, the helix or the electro-magnet, used to complete the circuit; N and P are the cross wires, which can either be brought into contact at *x*, or else have a galvanometer or an electrolyzing apparatus interposed there.

The production of the *shock* from the current in the cross wire, whether D was a long extended wire, or a helix, or an electromagnet, has been already described.

1084. *Chemical decomposition* was next effected by the cross-wire current, an electro-magnet being used at D, and a decomposing apparatus, with solution of iodide of potassium in paper, employed at *x*. The conducting power of the connecting system ABD was sufficient to carry all the primary current, and consequently no chemical action took place at *x* during the *continuance* of contact at G and E; but when contact was broken, there was instantly decomposition at *x*. The iodine appeared against the wire N, and not against the wire P; thus demonstrating that the current through the crosswires, when contact was broken, was in the *reverse direction* to that marked by the arrow, or that which the electromotor would have sent through it.

1087. The most instructive set of results was obtained, however, when the *galvanometer* was introduced at *x*. Using an electromagnet at D, and continuing contact, a current was then indicated by the deflection, proceeding from P to N, in the direction of the arrow; the cross-wire serving to carry one part of the electricity excited by the electromotor, and that part of the arrangement marked A B D, the other and far greater part, as indicated by the arrows. The magnetic needle was then forced back, by pins applied upon opposite sides of its two extremities, to its natural position when uninfluenced by a current; after which, contact being *broken* at G or E, it was deflected strongly in the opposite direction; thus showing, in accordance with the chemical effects, that the extra current followed a course in the cross-wires *contrary* to that indicated by the arrow, i. e. contrary to the one produced by the direct action of the electromotor.

1089. These experiments, establishing as they did, by the quantity, intensity, and even direction, a distinction between the primary or generating current and the extra current, led me to conclude that the latter was identical with the induced current described in the First Series of these Researches; and this opinion I was soon able to bring to proof, and at the same times obtained not the partial but entire separation of one current from the other.

1090. The double helix was arranged so that it should form the connecting wire between the plates of the electromotor, ii being out of the current, and its ends unconnected. In this condition i acted very well, and gave a good spark at the time and place of disjunction. The opposite ends of ii were then connected together so as to form an endless wire, i remaining unchanged: but now *no spark*, or one scarcely sensible, could be obtained from the latter at the place of disjunction. Then, again, the ends of ii were held so nearly together that any current running round that helix should be rendered visible as a spark; and in this manner a spark was obtained from ii when the junction of i with the electromotor was broken, in place of appearing at the disjoined extremity of i itself.

1091. By introducing a galvanometer or decomposing apparatus into the circuit formed by the helix ii, I could easily obtain the deflections and decomposition occasioned by the induced current due to the breaking contact at helix i, or even to that occasioned by making contact of that helix with the electromotor; the results in both cases indicating the contrary directions of the two induced currents thus produced.

1100. Thus all the phenomena tend to prove that the effects are due to an inductive action, occurring at the moment when the principal current is stopped. I at one time thought they were due to an action continued during the *whole time* of the current, and expected that a steel magnet would have an influence according to its position in the helix, comparable to that of a soft iron bar, in assisting the effect. This, however, is not the case; for hard steel, or a magnet in the helix, is not so effectual as soft iron; nor does it make any difference how the magnet is placed in the helix, and for very simple reasons, namely, that the effect does not depend upon a permanent state of the core, but a *change of state*; and that the magnet or hard steel cannot sink through such a difference of state as soft iron, at the moment contact ceases, and therefore cannot produce an equal effect in generating a current of electricity by induction.

1101. As an electric current acts by induction with equal energy at the moment of its commencement as at the moment of its cessation, but in a contrary direction, the reference of the effects under examination to an inductive action, would lead to the conclusion that corresponding effects of an opposite nature must occur in a long wire, a helix, or an electro-magnet, every time that *contact is made* with the electromotor. These effects will tend to establish a resistance for the first moment in the long conductor, producing a result equivalent to the reverse of a shock or a spark. Now it is very difficult to devise means fit for the recognition of such negative results; but as it is probable that some positive effect is produced at the time, if we knew what to expect, I think the few facts bearing upon this subject with which I am acquainted are worth recording.

1102. The electro-magnet was arranged with an electrolyzing apparatus at *x*, as before described, except that the intensity of the chemical action at the electromotor was increased until the electric current was just able to produce the feeblest signs of decomposition whilst contact was continued at G and E; (the iodine of course appearing against the end of the cross wire P;) the wire N was also separated from A at *r*, so that contact there could be made or broken at pleasure. Under these circumstances the following set of actions was repeated several times: contact was broken at *r*, then broken at G, next made at *r*, and lastly renewed at G; thus any current from N to P due to *breaking* of contact was avoided, but any additional force to the current from P to N due to *making* contact could be observed. In this way it was found, that a much greater decomposing effect (causing the evolution of iodine against P) could be obtained by a few completions of contact than by the current which could pass in a much longer time if the contact was *continued*. This I attribute to the act of induction in the wire A B D at the moment of contact rendering that wire a worse conductor, or rather retarding the passage of the electricity through it for the instant, and so throwing a greater quantity of the electricity which the electromotor could produce, through the cross wire passage N P. The instant the induction ceased, A B D resumed its full power of carrying a constant current of electricity, and could have it highly increased, as we know by the former experiments by the opposite inductive action brought into activity at the moment contact at Z or C was *broken*.

1103. A galvanometer was then introduced at *x*, and the deflection of the needle noted whilst contact was continued at G and E: the needle was then blocked as before in one direction, so that it should not return when the current ceased, but remain in the position in which the current could retain it. Contact at G or E was broken, producing of course no visible effect; it was then renewed, and the needle was instantly deflected, passing from the blocking pins to a position still further from its natural place than that which the constant current could give, and thus showing, by the temporary excess of current in this cross communication, the temporary retardation in the circuit A B D.

1104. On adjusting a platina wire at *x* so that it should not be ignited by the current passing through it whilst contact at G and E was *continued*, and yet become red-hot by a current somewhat more powerful, I was readily able to produce its ignition upon *making contact*, and again upon *breaking contact*. Thus the momentary retardation in A B D on making contact was again shown by this result, as well also as the opposite result upon breaking contact. The two ignitions of the wire at *x* were of course produced by electric currents moving in opposite directions.

1107. Thus the case, under the circumstances, is, that the intensity and quantity of electricity moving in a current are smaller when the current commences or is increased, and greater when it diminishes or ceases, than they would be if the inductive action occurring at these moments did not take place; or than they are in the original current wire if the inductive action be transferred from that wire to a collateral one.

1117. The effects produced at the commencement and end of a current, (which are separated by an interval of time when that current is supplied from a voltaic apparatus,) must occur at the same moment when a common electric discharge is passed through a long wire. Whether, if happening accurately at the same moment, they would entirely neutralize each other, or whether they would not still give some definite peculiarity to the discharge, is a matter remaining to be examined; but it is very probable that the peculiar character and pungency of sparks drawn from a long wire depend in part upon the increased intensity given at the termination of the discharge by the inductive action then occurring.

# Twenty-eighth Series[1]

## § 34. On lines of Magnetic Force; their definite character; and their distribution within a Magnet and through space

[Received October 22,—Read November 27 and December 11, 1851.]

3070. FROM my earliest experiments on the relation of electricity and magnetism, I have had to think and speak of lines of magnetic force as representations of the magnetic power; not merely in the points of quality and direction, but also in quantity. The necessity I was under of a more frequent use of the term in some recent researches, has led me to believe that the time has arrived, when the idea conveyed by the phrase should be stated very clearly, and should also be carefully examined, that it may be ascertained how far it may be truly applied in representing magnetic conditions and phænomena; how far it may be useful in their elucidation; and, also, how far it may assist in leading the mind correctly on to further conceptions of the physical nature of the force, and the recognition of the possible effects, either new or old, which may be produced by it.

3071. A line of magnetic force may be defined as that line which is described by a very small magnetic needle, when it is so moved in either direction correspondent to its length, that the needle is constantly a tangent to the line of motion; or it is that line along which, if a transverse wire be moved in either direction, there is no tendency to the formation of any current in the wire, whilst if moved in any other direction there is such a tendency; or it is that line which coincides with the direction of the magnecrystallic axis of a crystal of bismuth, which is carried in either direction along it. The direction of these lines about and

[1] Philosophical Transactions, 1852, p. 1.

amongst magnets and electric currents, is easily represented and understood, in a general manner, by the ordinary use of iron filings.

3072. These lines have not merely a determinate direction, recognizable as above, but because they are related to a polar or antithetical power, have opposite qualities or conditions in opposite directions; these qualities, which have to be distinguished and identified, are made manifest to us, either by the position of the ends of the magnetic needle, or by the direction of the current induced in the moving wire.

3073. A point equally important to the definition of these lines is, that they represent a determinate and unchanging amount of force. Though, therefore, their forms, as they exist between two or more centres or sources of magnetic power, may vary very greatly, and also the space through which they may be traced, yet the sum of power contained in any one section of a given portion of the lines is exactly equal to the sum of power in any other section of the same lines, however altered in form, or however convergent or divergent they may be at the second place. The experimental proof of this character of the lines will be given hereafter.

3074. Now it appears to me that these lines may be employed with great advantage to represent the nature, condition, direction and comparative amount of the magnetic forces; and that in many cases they have, to the physical reasoner at least, a superiority over that method which represents the forces as concentrated in centres of action, such as the poles of magnets or needles; or some other methods, as, for instance, that which considers north or south magnetisms as fluids diffused over the ends or amongst the particles of a bar. No doubt, any of these methods which does not assume too much, will, with a faithful application, give true results; and so they all ought to give the same results as far as they can respectively be applied. But some may, by their very nature, be applicable to a far greater extent, and give far more varied results, than others. For just as either geometry or analysis may be employed to solve correctly a particular problem, though one has far more power and capability, generally speaking, than the other; or just as either the idea of the reflexion of images, or that of the reverberation of sounds may be used to represent certain physical forces and conditions; so may the idea of the attractions and repulsions of centres, or that of the disposition of magnetic fluids, or that of lines of force, be applied in the consideration of magnetic phænomena. It is the occasional and more frequent use of the latter which I at present wish to advocate.

3075. I desire to restrict the meaning of the term *line of force*, so that it shall imply no more than the condition of the force in any given place, as to strength and direction; and not to include (at present) any idea of the nature of the physical cause of the phænomena; or be tied up with, or in any way dependent on, such an idea. Still, there is no impropriety in endeavouring to conceive the method in which the physical forces are either excited, or exist, or are transmitted; nor, when these by experiment and comparison are ascertained in any given degree, in representing them by any method which we adopt to represent the mere forces, provided no error is thereby introduced. On the contrary, when the natural truth and the conventional representation of it most closely agree, then are we most advanced in our knowledge. The emission and the æther theories present such cases in relation to light. The idea of a fluid or of two fluids is the same for electricity; and there the further idea of a current has been raised, which indeed has such hold on the mind as occasionally to embarrass the science as respects the true character of the physical agencies, and may be doing so, even now, to a degree which we at present little suspect. The same is the case with the idea of a magnetic fluid or fluids, or with the assumption of magnetic centres of action of which the resultants are at the poles. How the magnetic force is transferred through bodies or through space we know not:—whether the result is merely action at a distance, as in the case of gravity; or by some intermediate agency, as in the cases of light, heat, the electric current, and (as I believe) static electric action. The idea of magnetic fluids, as applied by some, or of magnetic centres of action, does not include that of the latter kind of transmission, but the idea of lines of force does. Nevertheless, because a particular method of representing the forces does not include such a mode of transmission, the latter is not therefore disproved; and that method of representation which harmonizes with it may be the most true to nature. The general conclusion of philosophers seems to be, that such cases are by far the most numerous, and for my own part, considering the relation of a vacuum to the magnetic force and the general character of magnetic phænomena external to the magnet, I am more inclined to the notion that in the transmission of the force there is such an action, external to the magnet, than that the effects are merely attraction and repulsion at a distance. Such an action may be a function of the æther; for it is not at all unlikely that, if there be an æther, it should have other uses than simply the conveyance of radiations. Perhaps when we are more clearly instructed in this matter, we shall see the source of the contradictions which are supposed to exist between the results of Coulomb, Harris and other philosophers, and find that they are not contradictions in reality, but mere differences in degree, dependent upon partial or imperfect views of the phænomena and their causes.

3076. Lines of magnetic force may be recognized, either by their action on a magnetic needle, or on a conducting body moving across them. Each of these actions may be employed also to indicate, either the direction of the line, or the force exerted at any given point in it, and this they do with advantages for the one method or the other under particular circumstances. The actions are however very different in their nature. The needle shows its results by attractions and repulsions; the moving conductor or wire shows it by the production of a current of electricity. The latter is an effect entirely unlike that produced on the needle, and due to a different action of the forces; so that it gives a view and a result of properties of the lines of force, such as the attractions and repulsions of the needle could never show. For this and other reasons I propose to develope and apply the method by a moving conductor on the present occasion.

3078. The moving wire produces its greatest effect and indication, not when passing from stronger to weaker places, or the reverse, but when moving in places of equal action, *i.e.* transversely across the lines of force.

3079. It determines the direction of the polarity by an effect entirely independent of pointing or suchlike results of attraction or repulsion; *i. e.* by the direction of the electric current produced in it during the motion.

3080. The principle can be applied to the examination of the forces *within* numerous solid bodies, as the metals, as well as outside in the air. It is not often embarrassed by the difference of the surrounding media, and can be used in fluids, gases or a vacuum with equal facility. Hence it can penetrate and be employed where the needle is forbidden; and in other cases where the needle might be resorted to, though greatly embarrassed by the media around it, the moving wire may be used with an immediate result.

3081. The method can even be applied with equal facility to the interior of a magnet, a place utterly inaccessible to the magnetic needle.

3082. The moving wire can be made to sum up or give the resultant at once of the magnetic action at many different places, *i. e.* the action due to an area or section of the lines of force, and so supply experimental comparisons which the needle could not give, except with very great labour, and then imperfectly. Whether the wire moves directly or obliquely across the lines of force, in one direction or another, it sums up, with the same accuracy in principle, the amount of the forces represented by the lines it has crossed.

3084. I will proceed to take the case of a simple bar magnet, employing it in illustration of what has been said respecting the lines of force and the moving conductor, and also for the purpose of ascertaining how these lines of force are disposed, both without and within the magnet itself, upon which they are dependent or to which they belong. For this purpose the following apparatus was employed. Let fig. 1 represent a wooden stand, of which the base is a board 17·5 inches in length, and 6 inches in breadth, and 0·8 of an inch in thickness: these dimensions will serve as a scale for the other parts. A B are two wooden uprights; D is an axis of wood having two long depressions cut into it, for the purpose of carrying the two bar magnets F and G. The wood is not cut away quite across the axis, but is left in the middle, so that the magnets are about $\frac{1}{15}$ of an inch apart. From O towards the support A, it is removed, however, as low down as the axis of revolution, so as to form a notch between the two magnets when they

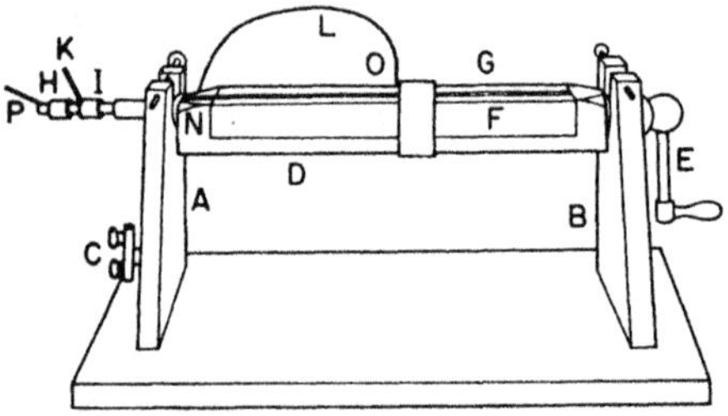


FIG. 1.

are in their places; and by further removal of the wood, this notch is continued on to the end of the axis at P. This notch, or opening, is intended to receive a wire, which can be carried down the axis of rotation, and then passing out between the two magnets, anywhere between O and N, can be returned towards the end P on the outside. The magnets are so placed, that the central line of their compound system coincides with the axis of rotation; E being a handle by which rotation, when required, is given. H and I are two copper rings, slipping tightly on to the axis, by which communication is to be made between a wire adjusted so as to revolve with the magnets, and the fixed ends of wires proceeding from a galvanometer. Thus, let P L represent a covered wire; which being led along the bottom of the notch in the axis of the apparatus, and passing out at the equatorial parts of the magnets, returns into the notch again near N, and terminates at K. When the form of the wire loop is determined and given to it, then a little piece of soft wood is placed between the wires in the notch at K, of such thickness, that when the ring I is put into its place, it shall press upon the upper wire, the piece of wood, and the lower wire, and keep all tightly fixed together, and at the same time leave the two wires effectually separated. The second ring, H, is then put into its place on the axis, and the introduction of a small wedge of wood, at the end of the axis, serves to press the end P into close and perfect contact with the ring H, and keep all in order. So the wire is free to revolve with the magnets, and the rings H and I are its virtual terminations. Two clips, as at C, hold the ends of the galvanometer wire (also of copper); and the latter are made to press against the rings by their elasticity, and give an effectual contact bearing, which generates no current, either by difference of nature or by friction, during the revolution of the axis.

3085. The two magnets are bars, each 12 inches long, 1 inch broad, and 0·4 of an inch thick. They weigh each 19 ounces, and are of such a strength as to lift each other end to end and no more. When the two are adjusted in their place, it is with the similar poles together, so that they shall act as one magnet, with a division down the middle: they are retained in their place by tying, or, at times, by a ring of copper which slips tightly over them and the axis.

3086. The galvanometer is a very delicate instrument made by Rhumkorff. It was placed about 6 feet from the magnet apparatus, and was not affected by any revolution of the latter. The wires, connecting it with the magnets, were of copper, 0·04 of an inch in diameter, and in their whole length about 25 feet. The length of the wire in the galvanometer I do not know; its diameter was $\frac{1}{135}$th of an inch. The condition of the galvanometer, wires, and magnets, was such, that when the bend of the wires was formed into a loop, and that carried once over the pole of the united magnets, as from *a* to *b*, fig. 2, the galvanometer needle was deflected two degrees or more. The vibration of the needle was slow, and it was easy therefore to reiterate this action five or six times, or oftener, breaking and making contact with the galvanometer at right intervals, so as to combine the effect of like induced currents; and then a deflection of 10° or 15° on either side of zero could be readily obtained. The arrangement, therefore, was sufficiently sensible for first experiments; and though the resistance opposed by the

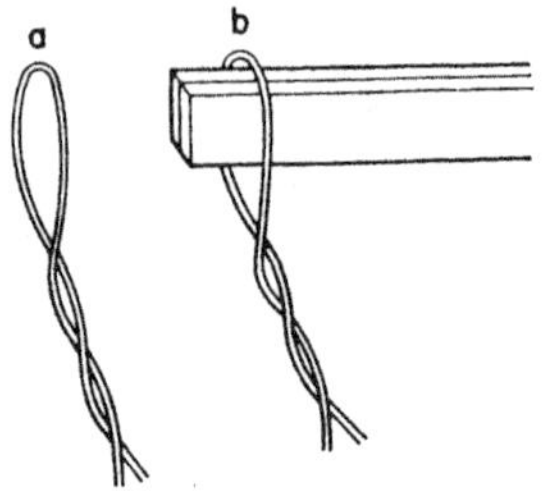


FIG. 2.

thin long galvanometer wire to feeble currents was considerable, yet it would always be the same, and would not interfere with results, where the final effect was equal to 0°, nor in those where the consequences were shown, not by absolute measurement, but by comparative differences.

3087. The first practical result produced by the apparatus described, in respect of magneto-electric induction generally, is, that a piece of metal or conducting matter which moves across lines of magnetic force, has, or tends to have, a current of electricity produced in it. A more restricted and precise expression of the full effect is the following. If a continuous circuit of conducting matter be traced out, or conceived of, either in a solid or fluid mass of metal or conducting matter, or in wires or bars of metal arranged in non-conducting matter or space; which being moved, crosses lines of magnetic force, or being still, is by the translation of a magnet crossed by such lines of force; and further, if, by inequality of angular motion, or by contrary motion of different parts of the circuit, or by inequality of the motion in the same direction, one part crosses either more or fewer lines than the other; then a current will exist round it, due to the differential relation of the two or more intersecting parts during the time of the motion: the direction of which current will be determined (with lines having a given direction of polarity) by the direction of the intersection, combined with the relative amount of the intersection in the two or more efficient and determining (or intersecting) parts of the circuit.

3091. In the first instance the wire was carried down the axis of the magnet to the middle distance, then led out at the equatorial part, and returned on the outside; fig. 4 will represent such a

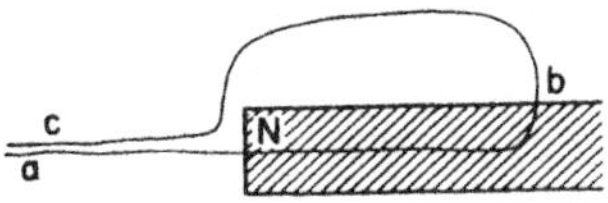


FIG. 4.

disposition. Supposing the magnet and wire to revolve once, it is evident that the wire *a* may be considered as passing in at the axis of the magnet, and returning from *b* across the lines of force external to the magnet, to the axis again at *c*; and that in one revolution, the wire from *b* to *c* has intersected once, all the lines of force emanating from the N end of the magnet. In other words, whatever course the wire may take from *b* to *c*, the whole system of lines belonging to the magnet has been *once* crossed by the wire. In order to have a correct notion of the relation of the result, we will suppose a person standing at the handle E, fig. 1, and looking along the magnets, the magnets being fixed, and the wire loop from *b* to *c* turned over toward the left-hand into a horizontal plane; then, if that loop be moved over towards the right-hand, the magnet remaining stationary, it will be equivalent to a *direct* revolution (according to the hands of a watch or clock) of 180°, and will produce a feeble current in a given direction at the galvanometer. If it be carried back 180° in the reverse direction, it will produce a corresponding current in the reverse direction to the former. If the wire be held in a vertical, or any other plane, so that it may be considered as fixed, and the magnet be rotated through half a revolution, it will also produce a current; and if rotated in the contrary direction, will produce a contrary current; but as to the *direction* of the currents, that produced by the *direct* revolution of the wire is the same as that produced by the *reverse* revolution of the magnet; and that produced by the *reverse* revolution of the wire is the same as that produced by the *direct* revolution of the magnet. A more precise reference of the direction of the current to the particular pole employed, and the direction of the revolution of the wire or magnet, is not at present necessary; but if required is obtained at once by reference to the general law.

3092. The magnet and loop being rotated together in either direction, no trace of an electric current was produced. In this case the effect, if any, could be greatly exalted, because the rotation could be continued for 10, 20, or any number of revolutions without derangement, and it was easy to make thirty revolutions or more within the time of the swing of the galvanometer needle in one direction. It was also easy, if any effect were produced, to accumulate it upon the galvanometer by reversing the rotation at the due time. But no amount of revolution of the magnet and wire together could produce any effect.

3093. The loop was then taken out of the axis of the magnet, but attached to it by a piece of pasteboard, so that all should be fixed together and revolve with the same angular velocity, fig. 5;

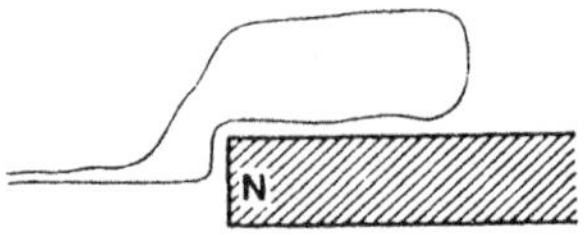


FIG. 5.

but whatever the shape or disposition of the loop, whether large or small, near or distant, open or shut, in one plane, or contorted into various planes; whatever the shape or condition, or place, provided it moved altogether with the magnet, no current was produced.

3094. Furthermore, when the loop was out of the magnets, and by expedients of arrangement, was retained immoveable, whilst the magnet revolved, no amount of rotation of the magnet (unaccompanied by translation of place) produced any degree of current through the loop.

3095. The loop of wire was then made of two parts; the portion *c*, fig. 6, on the outside of the magnet, was fixed at *b*, and the

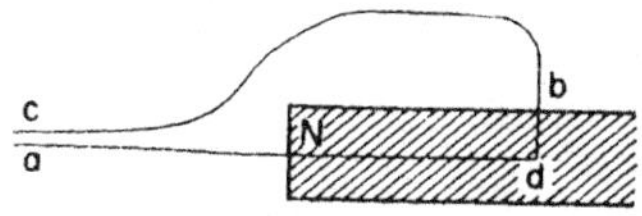


FIG. 6.

portion *a*, being a separate piece, was carried along the axis until it came in contact with the former at *d*; the revolution of one part was thus permitted either with or without the other, yet preserving always metallic contact and a complete circuit for the induced current. In this case, when the external wire and the magnet were fixed, no current was produced by any amount of revolution of the wire *a* on its axis. Neither was any current produced when the magnet and wire, *c d*, were revolved together, whether the wire *a* revolved with them or not. When the magnet was revolved without the external part of *c d*, or the latter revolved without the magnet, then currents were produced as before.

3096. The magnet was now included in the circuit, in the following manner. The wire *a*, fig. 7, was placed in metallic contact on

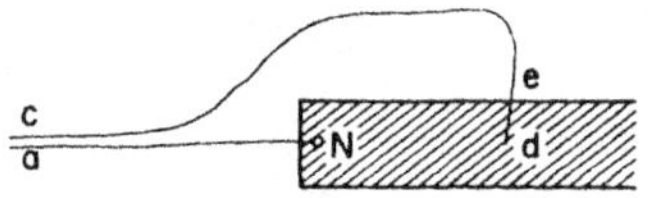


FIG. 7.

both sides of the interval between the magnets at N (or the pole), and the part *c* was brought into contact with the centre at *d*. The result was in everything the same as when the wire *a* was continued up to *d*, *i.e.* no amount of revolution of the magnet and part *c* together could produce any electric current. When *c* was made to terminate at *e* or the equatorial part of the magnet, the result was precisely the same. Also, when *c* terminated at *e*, the part *a* of the wire was continued to the centre at *d*, and there the contact perfected, but the result was still the same. No difference, therefore, was produced, by the use between N and *d*, or *d* and *e*, of the parts of the magnet in place of an insulated copper wire, for the completion of the circuit in which the induced current was to travel. No rotation of the part *a* produced any effect, wherever it was made to terminate.

3097. In order to obtain the power of rotating the magnet without the external part of the wire, a copper ring was fixed round, and in contact with it at the equatorial part, and the wire *c*, fig. 8,

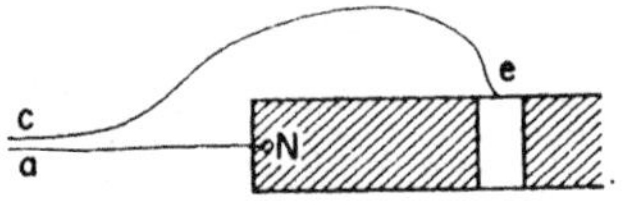


FIG. 8.

made to bear by spring pressure against this ring, and also against the ring H on the axis, fig. 1; the circuit was examined, and found complete. Now when the wire *c e* was fixed and the magnet rotated, a current was produced, and that to the same amount for the same number of revolutions, whether the part of the wire *a* terminated at N, or was continued on to the centre of the magnet, or was insulated from the magnet and continued up to the copper ring *e*. When the wire, by expedients, which though rough were sufficient, was made to revolve whilst the magnet was still, currents in the contrary direction were produced, in accordance with the effect before described; and the results when the wire and magnet rotated together, show that these are in amount exactly equal to the former. When the inner and the outer wires were both motionless, and the magnet only revolved, a current in the full proportion was produced, and that, whether the axial wire *a* made contact at the pole of the magnet or in the centre.

3098. Another arrangement of the magnet and wires was of the following kind. A radial insulated wire was fixed in the middle of the magnets, from the centre *d*, fig. 9, to the circumference *b*, being connected there with the equatorial ring; an axial wire touched this radial wire at the centre and passed out at the pole;

the external part of the circuit, pressing on the ring at the equator, proceeded on the outside over the pole to form the communication as before. In the case where the magnet was revolved without the

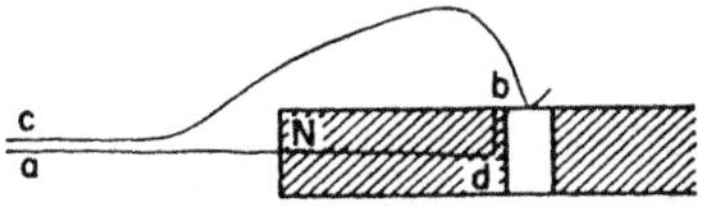


FIG. 9.

axial and the external wire, the full and proper current was produced; the small wire, *d b*, being, however, the only part in which this current could be generated by the motion; for it replaced, under these circumstances, the body of the magnet employed on the former occasion.

3099. The external part of the wire instead of being carried back over that pole of the magnet at which the axial wire entered, was continued away over the other pole, and so round by a long circuit to the galvanometer; still the revolution of the magnet, under any of the described circumstances, produced exactly the same results as before. It will be evident by inspection of fig. 10,

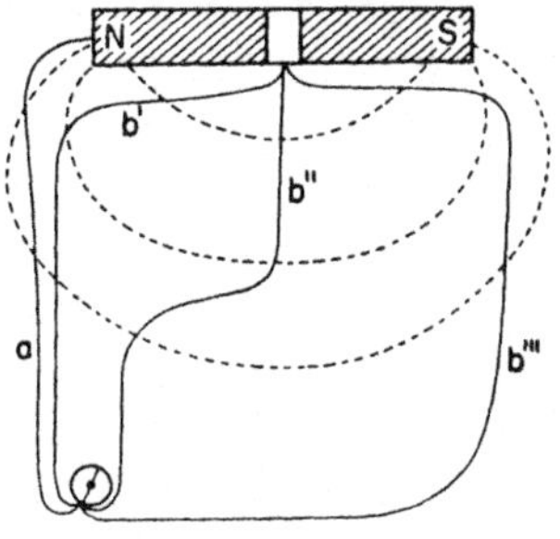


FIG. 10.

that, however the wires are carried away, the general result will, according to the assumed principles of action, be the same; for if *a* be the axial wire, and *b′*, *b″*, *b‴* the equatorial wire, represented in three different positions, whatever magnetic lines of force pass across the latter wire in one position, will also pass across it in the other, or in *any* other position which can be given to it. The distance of the wire at the place of intersection with the lines of force, has been shown, by the experiments, to be unimportant.

3100. Whilst considering the condition of the forces of a magnet, it may be admitted, that the two magnets used in the experimental investigations described, act truly as one central magnet. We have only to conceive smaller similar magnets to be introduced to fill up the narrow space not occupied by the wire, and then the complete magnet would be realized:—or it may be viewed as a magnet once perfect, which has had certain parts removed; and we know that neither of these changes would disturb the general disposition of the forces. In and around the bar magnet the forces are distributed in the simplest and most regular manner. Supposing the bar removed from other magnetic influences, then its power must be considered as extending to any distance, according to the recognized law; but, adopting the representative idea of *lines of force*, any wire or line proceeding from a point in the magnetic equator of the bar, over one of the poles, so as to pass through the magnetic axis, and so on to a point on the opposite side of the magnetic equator, must intersect *all* the lines in the plane through which it passes, whether its course be over the one pole or the other. So also a wire proceeding from the end of the magnet at the magnetic axis, to a point at the magnetic equator, must intersect curves equal to half those of a great plane, however small or great the length of the wire may be; and though by its tortuous course it may pass out of one plane into another on its way to the equator.

3101. Further, if such a wire as that last described be revolved once round the end of the magnet to which it is related, a slipping contact at the equator being permitted for the purpose, it will intersect *all* the lines of force during the revolution; and that, whether the polar contact is absolutely coincident with the magnetic axis, or is anywhere else at the end of the bar, provided it remain for the time unchanged. All this is true, though the magnet may be subject, by induction at a distance, to other magnets or bodies, and may be exerting part of its force on them, so as to make the distribution of its power very irregular as compared to the case of the independent bar, or may have an irregular or contorted shape, even up to the horseshoe form. It is evident, indeed, that if a wire have one of its ends applied to *any* point on the surface of a magnet, and the other end to a point in the magnetic equator, and the latter be slipped once round the magnetic equator, and the loop of wire be made to pass over either pole, so as at last to resume its first position, it will in the course of its journey have intersected *once* every line of force belonging to the magnet.

3107. Proceeding to experiment upon the effect of the *distance* of the wire *c*, fig. 11, from the magnet, the wire was made to vary,

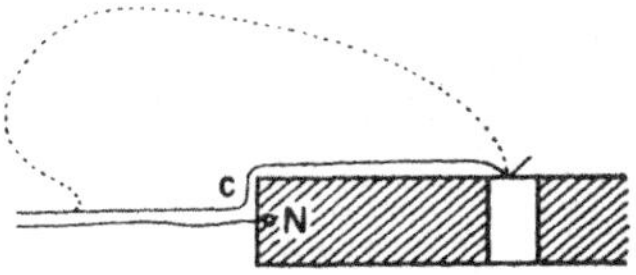


FIG. 11.

so that sometimes it was not more than 8 inches long (being of copper and 0·04 of an inch in diameter), and only half an inch from the magnet, whilst at other times it was 6 or 8 feet long and extended to a great distance. The deflection due to ten revolutions of the magnet was observed, and the average of several observations, for each position of the wire, taken: these were very close (with the precautions before described) for the same position; and the averages for different positions agreed perfectly together, being 9°·5. I endeavoured to repeat these experiments on distance by moving the wire and preserving the magnet stationary in the manner before described; they were not so striking because time would only allow of smaller deflections being obtained, but the same number of journeys through an arc of 180° gave the same deflection at the galvanometer, whether the course of the wire was close to the magnet or far off; and the deflection agreed with those obtained when the magnet was rotating and the wire at rest.

3109. From these results the following conclusions may be

drawn. The *amount* of magnetic force, as shown by its effect in evolving electric currents, is determinate for the same lines of force, whatever the distance of the point or plane, at which their power is exerted, is from the magnet. Or it is the same in any two, or more, sections of the same lines of force, whatever their form or their distance from the seat of the power may be. This is shown by the results with the magnet and the wire, when both are in the circuit; and also by the wire loop revolving with the magnet; where the tendency of currents to form in the two parts oppose and exactly neutralize or compensate each other.

3110. In the latter case very varying sections outside of the magnet may be compared to each other; thus, the wire may be conceived of as passing (or be actually formed so as to intersect) lines of force near the pole, and then, being continued *along* a line of force until over the equator, may be directed so as to intersect the same lines of force in the contrary direction, and then return along a line of force to its commencement; and so two surface sections may be compared. It is manifest that every loop forming a complete circuit, which is in a great plane passing through the axis of the magnet, must have precisely the same lines of force passing into and passing out of it, though they may, so to say, be expanded in one part and compressed in another; or (speaking in the language of radiation) be more intense in one part and less intense in the other. It is also as manifest, that, if the loop be not in one plane, still, on making one complete revolution, either with or without the magnet, it will have intersected in its two opposite parts an exactly equal amount of lines of force. Hence the comparison of any one section of a given amount of lines of force with any other section is rendered, experimentally, very extensive.

3111. Such results prove, that, under the circumstances, there is no loss, or destruction, or evanescence, or latent state of the magnetic power by distance.

3112. Also that convergence or divergence of the lines of force causes no difference in their amount.

3113. That obliquity of intersection causes no difference. It is easy so to shape the loop, that it shall intersect the lines of force directly across at both places of intersection, or directly at one and obliquely at the other, or obliquely in any degree at both; and yet the result is always the same.

3114. It is also evident, by the results of the rotation of the wire and magnet, that when a wire is moving amongst equal lines (or in a field of equal magnetic force), and with an uniform motion, then the current of electricity produced is proportionate to the *time*; and also to the *velocity* of motion.

3115. They also prove, generally, that the quantity of electricity thrown into a current is directly as the amount of curves intersected.

3122. On the other hand, the use of the idea of *lines of force*, which I recommend, to represent the true and real magnetic forces, makes it very desirable that we should find a unit of such force, if it can be attainable, by any experimental arrangement, just as one desires to have a unit for rays of light or heat. It does not seem to me improbable that further research will supply the means of establishing a standard of this kind. In the meantime, for the enlargement of the utility of the idea in relation to the magnetic force, and to indicate its conditions graphically, lines may be employed as representing these units in any given case. I have so employed them in former series of these Researches, where the direction of the *line of force* is shown at once, and the relative amount of force, or of lines of force in a given space, indicated by their concentration or separation, *i.e.* by their number in that space. Such a use of unit lines involves, I believe, no error either in the direction of the polarity or in the amount of force indicated at any given spot included in the diagrams.

3171. The relation of the induced current to the electro-conducting power of the substance, amongst the metals, leads to the presumption that with other bodies, as water, wax, glass, &c., it is absent, only in consequence of the great deficiency of conducting power. I thought that processes analogous to those employed with the metals, might in such non-conductors as shell-lac, sulphur, &c., yield some results of static electricity; and have made many experiments with this view in the intense magnetic field, but without any distinct result.

3172. All the results described are those obtained with *moving metals*. But mere motion would not generate a relation, which had not a foundation in the existence of some previous state; and therefore the *quiescent* metals must be in some relation to the active centre of force, and that not necessarily dependent on their paramagnetic or diamagnetic condition, because a metal at zero in that respect, would have an electric current generated in it as well as the others. The relation is not as the attractions or repulsions of the metals, and therefore not magnetic in the common sense of the word; but according to some other function of the power. Iron, copper, and bismuth are very different in the former sense, but when moving across the lines of force, give the same general result, modified only by electro-conducting power.

3173. If such a condition be hereafter verified by experiment and the idea of an electrotonic state be revived and established, then, such bodies as water, oil, resin, &c., will probably be included in the same state; for the non-conducting condition, which prevents the formation of a current in them, does not militate against the existence of that condition which is prior to the effect of motion. A piece of copper, which cannot have the current, because it is not in a circuit, and a piece of lac, which cannot, because it is a non-conductor of electricity, may have peculiar but analogous states when moving across a field of magnetic power.

3174. On bringing this paper to a close, I cannot refrain from again expressing my conviction of the truthfulness of the representation, which the idea of lines of force affords in regard to magnetic action. All the points which are experimentally established with regard to that action, *i. e.* all that is not hypothetical, appear to be well and truly represented by it. Whatever idea we employ to represent the power, ought ultimately to include electric forces, for the two are so related that one expression ought to serve for both. In this respect, the idea of lines of force appears to me to have advantages over the method of representing magnetic forces by centres of action. In a straight wire, for instance, carrying an electric current, it is apparently impossible to represent the magnetic forces by centres of action, whereas the lines of force simply and truly represent them. The study of these lines have, at different times, been greatly influential in leading me to various results, which I think prove their utility as well as fertility. I have been so accustomed, indeed, to employ them, and especially in my last Researches, that I may unwittingly, have become prejudiced in their favour, and ceased to be a clear-sighted judge. Still, I have always endeavoured to make experiment the test and controller of theory and opinion; but neither by that nor by close cross examination in principle, have I been made aware of any error involved in their use.

# Twenty-ninth Series[1]

## § 35. On the employment of the Induced Magneto-electric Current as a test and measure of Magnetic Forces

[Received December 31, 1851,—Read March 25 and April 1, 1852.]

3177. THE proposition which I have made to use the induced magneto-electric current as an experimental indication of the presence, direction and amount of magnetic forces, makes it requisite that I should also clearly demonstrate the principles and develope the practice necessary for such a purpose; and especially that I should prove that the amount of current induced is precisely proportionate to the amount of lines of magnetic force intersected by the moving wire, in which the electric current is generated and appears. The proof already given is, I think, sufficient for those who may repeat the experiments; but in order to accumulate evidence, as is indeed but proper in the first announcement of such a proposition, I proceeded to experiment with the magnetic power of the earth, which presents us with a field of action, not rapidly varying in force with the distance, as in the case of small magnets, but one which for a given place may be considered as uniform in power and direction; for if a room be cleared of all common magnets, then the terrestrial lines of magnetic force which pass through it, have one common direction, being that of the dip, as indicated by a free needle or other means, and are in every part in equal proportion or quantity, *i.e.* have equal power. Now the force being the same everywhere, the proportion of it to the current evolved in the moving wire is then perhaps more simply and directly determined, than in the case where, a small magnet being

[1] Philosophical Transactions, 1852, p. 137.

employed, the force rapidly changes in amount with the distance. . . .

## Revolving Rectangles and Rings

3192. The form of the moving wire which I have adopted for experiments with the magnetic forces of the earth, is either that of a rectangle or a ring. If a wire rectangle (fig. 3) be placed in a

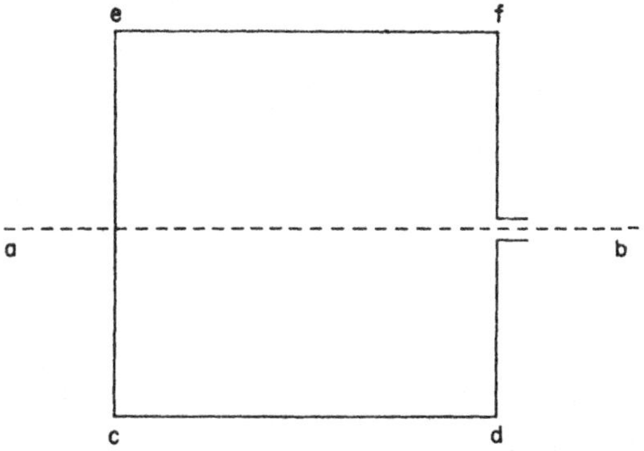


FIG. 3.

plane, perpendicular to the dip and then turned once round the axis *ab*, the two parts *cd* and *ef* will twice intersect the lines of magnetic force within the area *cedf*. In the first 180° of revolution the contrary direction in which the two parts *cd* and *ef* intersect those lines, will cause them to conspire in producing one current, tending to run round the rectangle in a given direction; in the following 180° of revolution they will combine in their effect to produce a contrary current; so that if the first current is from *d* by *ce* and *f* to *d* again, the second will be from *d* by *fe* and *c* to *d*. If the rectangle, instead of being closed, be open at *b*, and the ends there produced be connected with a commutator, which changes sides when the rectangle comes into the plane perpendicular to the dip, *i. e.* at every half revolution, then these successive currents can be gathered up and sent on to the galvanometer to be measured. The parts *ce* and *df* of the rectangle may be looked upon simply as conductors; for as they do not in their motion intersect any of the lines of force, so they do not tend to produce any current.

3193. The apparatus which carries these rectangles, and is also the commutator for changing the induced currents, consists of two uprights, fixed on a wooden stand, and carrying above a wooden horizontal axle, one end of which is furnished with a handle, whilst the other projects, and is shaped as in fig. 4. It

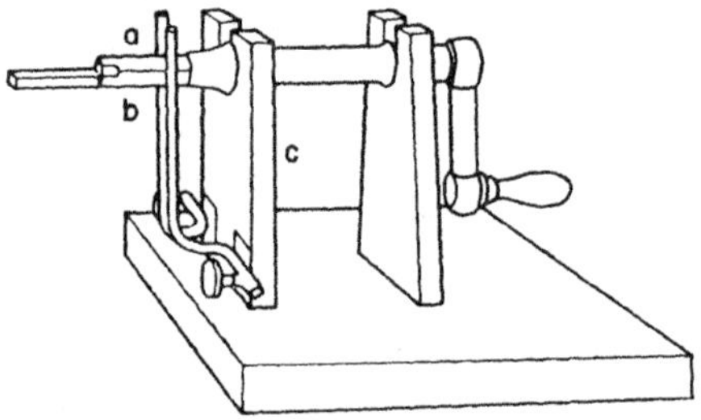


FIG. 4.

may there be seen, that two semicylindrical plates of copper *ab* are fixed on the axle, forming a cylinder round it, except that they do not touch each other at their edges, which therefore leave two lines of separation on opposite sides of the axle. Two strong copper rods, 0·2 of an inch in diameter, are fixed to the lower part of the upright *c*, terminating there in sockets with screws for the purpose of receiving the ends of the rods proceeding from the galvanometer cups: in the other direction the rods rise up parallel to each other, and being perfectly straight, press strongly against the curved plates of the commutator on opposite sides; the consequence is, that, whenever in the rotation of the axle, the lines of separation between the commutator plates arrive at and pass the horizontal plane, their contact with these bearing rods is changed, and consequently the direction of the current proceeding from these plates to the rods, and so on to the galvanometer, is changed also. The other or outer ends of the commutator plates are tinned, for the purpose of being connected by soldering to the ends of any rectangle or ring which is to be subjected to experiment.

3194. The rectangle itself is tied on to a slight wooden cross (fig. 5), which has a socket on one arm that slides on to and over

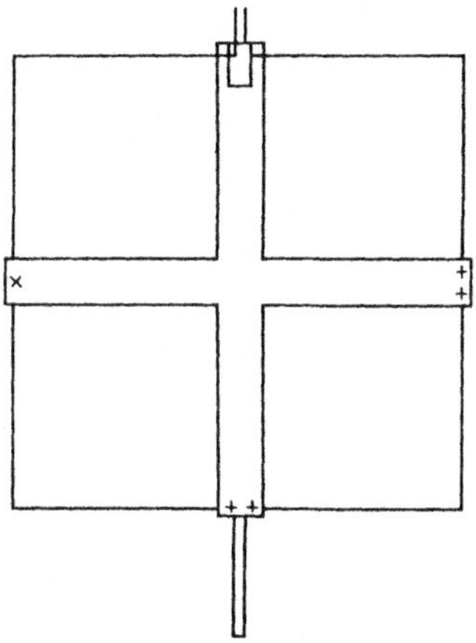

FIG. 5.

part of the wooden axle projecting beyond the commutator plates, so that it shall revolve with the axle. A small copper rod forms a continuation of that part of the frame which occupies the place of the axle, and the end of this rod enters into a hole in a separate upright, serving to support and steady the rectangle and its frame. The frames are of two or three sizes, so as to receive rectangles of 12 inches in the side, or even larger, up to 36 inches square. The rectangle is adjusted in its place, so that it shall be in the horizontal plane when the division between the commutator plates is in the same plane, and then its extremities are soldered to the two commutator plates, one to each. It is now evident, that when dealing with the lines of force of the earth, or any other lines, the axle has only to be turned until the upright copper rods touch on each side at the separation of the commutator plates, and then the instrument adjusted in position, so that the plane of the ring or rectangle is perpendicular to the direction of the lines of force which are to be examined, and then any revolution of the commutator and intersecting wire will produce the maximum current which such wire and such magnetic force can produce. The lines of terrestrial magnetic force are inclined at an angle of 69° to the horizontal plane. As, however, only comparative results were required, the instrument was, in all the ensuing experiments, placed in the horizontal plane, with the axis of rotation perpendicular to the plane of the magnetic meridian; under which circumstances no cause of error or variation was introduced into the results. As no extra magnet was employed, the commutator was placed within three feet of the galvanometer, so that the two pieces of copper 3 feet long and 0·2 of an inch in thickness, sufficed to complete the communication. One end of each of these dipped into the galvanometer mercury cups, the other end was tinned, amalgamated, introduced into the sockets of the commutator rods, and secured by the pinching screw (fig. 4).

3195. When a given length of wire is to be disposed of in the form best suited to produce the maximum effect, then the circumstances to be considered are contrary for the case of a loop to be employed with a small magnet, and a rectangle or other formed loop to be employed with the lines of terrestrial force. In the case a small magnet, *all* the lines of force belonging to it are inclosed by the loop; and if the wire is so long that it can be formed into a loop of two or more convolutions, and yet pass over the pole, then twice or many times the electricity will be evolved that a single loop can produce. In the case of the earth's force, the contrary result is true; for as in circles, squares, similar rectangles etc. the areas inclosed are as the squares of the periphery, and the lines of force intersected are as the areas, it is much better to arrange a given wire in one simple circuit than in two or more convolutions. Twelve feet of wire in one square intersects in one revolution the lines of force passing through an area of nine square feet, whilst if arranged in a triple circuit, about a square of one foot area, it will only intersect the lines due to that area; and it is thrice as advantageous to intersect the lines within nine square feet once, as it is to intersect those of one square foot three times.

3196. A square was prepared, containing 4 feet in length of copper wire 0·05 of an inch in diameter; it enclosed one square foot of area, and was mounted on the commutator and connected in the manner already described. Six revolutions of it produced a swing deflection of 14° or 15°, and twelve quick revolutions were possible in the required time. The results of *quick* and *slow* revolutions were first compared. Six slow revolutions gave as the average of several experiments 15°·5 swing. Six moderate revolutions gave also an average of 15°·5; six quick revolutions gave an average of 15°·66. At another time twelve moderate revolutions gave an average of 28°·75, and twelve quick revolutions gave an average of 31°·33 swing. As before explained, the probable reason why the quick revolutions gave a larger result than the moderate or slow revolutions is, that in slow time the later revolutions are performed at a period when the needle is so far from parallel with the copper coil of the galvanometer that the impulses due to them are less effectually exerted. Hence a small or moderate number of revolutions and a quick motion is best. The difference in the extreme case is less than might have been expected, and shows that there is no practical objection in this respect to the method proposed of experimenting with the lines of magnetic force.

3197. In order to obtain for the present an expression of the power of the earth's magnetic force by this rectangle, observations were made on both sides of zero, as already recommended. Nine moderately quick direct revolutions (*i. e.* as the hands of a clock) gave as the average of many experiments 23°·87, and nine reverse revolutions gave 23°·37; the mean of these is 23°·62 for the nine revolutions of the rectangle, and therefore 2°·624 per revolution. Now the six quick revolutions gave 15°·66, which is 2°·61 per revolution, and the twelve quick revolutions gave 31°·33, which is also 2°·61 per revolution; and these results of 2°·624, 2°·61, and

2°·61 are very much in accordance, and give great confidence in this method of investigating magnetic forces.[1]

3198. A rectangle was prepared of the same length (4 feet) of the same wire, but the sides were respectively 8 and 16 inches (fig. 6), so that when revolving the intersecting parts should be only 8 inches in length instead of 12. The area of the rectangle was necessarily 128 square inches instead of 144. This rectangle showed the same difference of quick and slow rotation as before.

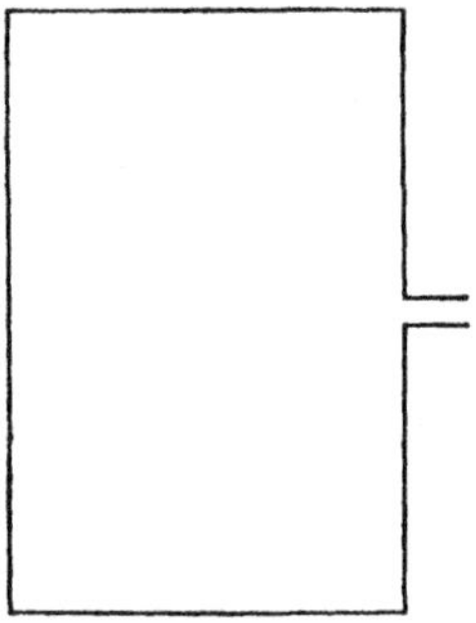

FIG. 6.

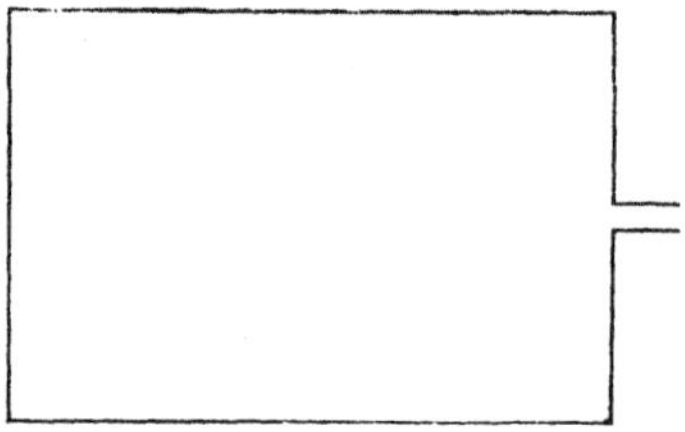

FIG. 7.

When nine direct revolutions were made, the result was 20°·87 swing. Nine reverse revolutions gave an average of 20°·25 swing: the mean is 20°·56, or 2°·284 per revolution. A third rectangle was prepared of the same length and kind of wire, the sides of which were respectively 8 inches and 16 inches long (fig. 7), but now so revolved that the intersecting parts were 16 inches, or twice as long as before; the area of the rectangle remained the same, *i. e.* 128 inches. The like effect of slow and quick revolutions appeared as in the former cases. Nine direct revolutions gave as the average

[1]
$$\sin\frac{15\cdot 66}{2} = \sin 7^\circ\cdot 83 = \sin 7^\circ\, 49'\cdot 8 = \cdot 1362343 \quad \frac{1362343}{6} = \cdot 0227057$$

$$\sin\frac{23\cdot 62}{2} = \sin 11\cdot 81 = \sin 11\; 48\cdot 6 = \cdot 2047069 \quad \frac{2047069}{9} = \cdot 0227474$$

$$\sin\frac{31\cdot 33}{2} = \sin 15\cdot 665 = \sin 15\; 40 = \cdot 2700403 \quad \frac{2700403}{12} = \cdot 0225034$$

effect 20°·75; and nine reverse revolutions produced 21°·375; the mean is 21°·06, or 2°·34 per revolution.

3199. Now 2°·34 is so near to 2°·284, that they may in the present state of the investigation be considered the same. The little difference that is evident, was, I suspect, occasioned by centrifugal power throwing out the middle of the longer intersecting parts during the revolution. The coincidence of the numbers shows, that the variation in the arrangement of the rectangle and in the length of the parts of the wires intersecting the lines of magnetic force, have had no influence in altering the result, which, being dependent alone on the number of lines of force intersected, is the same for both; for the area of the rectangles is the same. This is still further shown by comparing the results with those obtained with the square. The area in that case was 144 square inches, and the effect per revolution 2°·61. With the long rectangles the area is 128 square inches, and the mean of the two results is 2°·312 per revolution. Now 144 square inches is to 128 square inches as 2°·61 is to 2°·32; a result so near to 2°·312 that it may be here considered as the same; proving that the electric current induced is directly as the lines of magnetic force intersected by the moving wire.

3208. A large square was now constructed of copper wire 0·2 of an inch in diameter. The square was 36 inches in the side, and therefore consisted of 12 feet of wire, and enclosed an area of 9 square feet; it was attached to the commutator by expedients, which though sufficient for the present, were not accurate in the adjustments. It produced a fine effect upon the thick wire galvanometer; for one revolution caused a swing deflection of 80° or more; and when its rotation was continuous the needles were permanently deflected 40° or 50°. It was very interesting to see how, when this rectangle commenced its motion from the horizontal plane, the current increased in its intensity and then diminished again, the needles showing, that whilst the first 10° or 20° of revolution were being passed, there was very little power exerted on them; but that when it was towards or near the 90°, the power was great; the wires then intersecting the lines of force nearly at right angles, and therefore, with an equal velocity, crossing the greatest number in a given time. It was also very interesting, by the same indications, to see the two chief impulses given in one revolution of the rectangle. Being large and massive in proportion to the former wires, more time was required for a rotation than before, and the point of *time* or *velocity* of rotation became more essential. One rotation in a second was as much as I could well produce. A speed somewhat less than this was easy, convenient and quick enough; it gave for a single revolution near 80°, whilst a revolution with one-half or one-third the velocity, or less, gave only 60°, 50°, or even smaller amounts of deflection.

3209. Observations were now made on the measurement of one rotation having an easy quick velocity. The average of fifteen observations to the right, which came very close to each other was 78°·846; the average of seventeen similar observations to the left was 78°·382; and the mean of these results, or 78°·614, I believe to be a good first expression for this rectangle. On measuring the distance across after this result, I found that in one direction, *i. e.* across between the intersecting portions of wire, it was rather less than 36 inches; having therefore corrected this error, I repeated the observations and obtained the result of 81°·44. The difference of 2°·83, I believe to be a true result of the alteration and increase in area on making it more accurately 9 square feet; and it is to me an evidence of the sensibility and certainty of the instrument.

3210. As the two impulses upon the needles in one revolution are here sensibly apart in time, and as the needle has as evidently and necessarily left its first place before the second impulse is impressed upon it, so, that second impulse cannot be so effectual as the first. I therefore observed the results with half a revolution, and obtained a mean of 41°·37 for the effect. This number evidently belongs to the first of the two impulses of one revolution; and if we subtract it from 81°·44, it gives 40°·07 as the value of the second impulse under the changed place of the needle. This difference of the two impulses of one revolution namely 41°·37 and 40°·07, is in perfect accordance with the results that were to be expected.

3211. The square of this same copper wire, 0·2 inch in thickness, employed on a former occasion, had an area of one square foot, so that then the lines of force affected or affecting the moving wire, were one-ninth part of what they are in the present case: the effect then was 8°·94 per revolution. If, in comparing these cases, we take the ninth part of 81°·44, it gives 9°·04; a number so near the former, that we may consider the two rectangles as proving the same result, and at the same time the truth of the statement, that the magneto-electric current evolved is as the amount of lines of force intersected. A ninth part of the result with the large rectangle (78°·614), before its area was corrected, is 8°·734; so that the one is above and the other below the amount of the 12-inch rectangle. As that was not very carefully adjusted, nor indeed in any of the arrangements made as yet with extreme accuracy, I have little doubt that with accurately adjusted rectangles the results would be strictly proportional to the areas.

# On the Physical Lines of Magnetic Force

[Royal Institution Proceedings, June 11, 1852].

ON a former occasion certain lines about a bar-magnet were described and defined (being those which are depicted to the eye by the use of iron filings sprinkled in the neighbourhood of the magnet), and were recommended as expressing accurately the nature, condition, direction, and amount of the force in any given region either within or outside of the bar. At that time the lines were considered in the abstract. Without departing from or unsettling anything then said, the inquiry is now entered upon of the possible and probable *physical existence* of such lines. Those who wish to reconsider the different points belonging to these parts of magnetic science may refer to two papers in the first part of the Phil. Trans. for 1852 for data concerning the *representative* lines of force, and to a paper in the Phil. Mag. 4th Series, 1852, vol. iii. p. 401, for the argument respecting the *physical* lines of force.

Many powers act manifestly at a distance; their physical nature is incomprehensible to us: still we may learn much that is real and positive about them, and amongst other things something of the condition of the space between the body acting and that acted upon, or between the two mutually acting bodies. Such powers are presented to us by the phænomena of gravity, light, electricity, magnetism, &c. These when examined will be found to present remarkable differences in relation to their respective lines of forces; and at the same time that they establish the existence of real physical lines in some cases, will facilitate the consideration of the question as applied especially to magnetism.

When two bodies, *a*, *b*, gravitate towards each other, the line in which they act is a straight line, for such is the line which either would follow if free to move. The attractive force is not altered, either in *direction* or *amount*, if a third body is made to act by gravitation or otherwise upon either or both of the two first. A balanced cylinder of brass gravitates to the earth with a weight exactly the same, whether it is left like a pendulum freely to hang towards it, or whether it is drawn aside by other attractions or by tension, whatever the amount of the latter may be. A new gravitating force may be exerted upon *a*, but that does not in the least affect the amount of power which it exerts towards *b*. We have no evidence that *time* enters in any way into the exercise of this power, whatever the distance between the acting bodies, as that from the sun to the earth, or from star to star. We can hardly conceive of this force in one particle by itself; it is when two or more are present that we comprehend it: yet in gaining this idea we perceive no difference in the character of the power in the different particles; all of the same kind are *equal*, *mutual*, and *alike*. In the case of gravitation, no effect which sustains the idea of an independent or physical line of force is presented to us; and as far as we at present know, the line of gravitation is merely an ideal line representing the direction in which the power is exerted.

Take the Sun in relation to another force which it exerts upon the earth, namely its illuminating or warming power. In this case rays (which are lines of force) pass across the intermediate space; but then we may affect these lines by different media applied to them in their course. We may alter their direction either by reflection or refraction; we may make them pursue curved or angular courses. We may cut them off at their origin and then search for and find them before they have attained their object. They have a relation to *time*, and occupy 8 minutes in coming from the sun to the earth: so that they may exist independently either of their source or their final home, and have in fact a clear distinct physical existence. They are in extreme contrast with the lines of gravitating power in this respect; as they are also in respect of their condition at their terminations. The two bodies terminating a line of gravitating force are alike in their actions in every respect, and so the line joining them has like relations in both directions. The two bodies at the terminals of a ray are utterly unlike in action; one is a source, the other a destroyer of the line; and the line itself has the relation of a stream flowing in one direction. In these two cases of gravity and radiation, the difference between an abstract and a physical line of force is immediately manifest.

Turning to the case of Static Electricity we find here attractions (and other actions) at a distance as in the former cases; but when we come to compare the attraction with that of gravity, very striking distinctions are presented which immediately affect the question of a physical line of force. In the first place, when we examine the bodies bounding or terminating the lines of attraction, we find them as before, mutually and equally concerned in the action; but they are not alike: on the contrary, though each is endued with a force which speaking generally is of the like nature, still they are in such contrast that their actions on a third body in a state like either of them are precisely the reverse of each other,—what the one attracts the other repels; and the force makes itself evident as one of those manifestations of power endued with a dual and antithetical condition. Now with all such dual powers, attraction cannot occur unless the two conditions of force are present and in face of each other through the lines of force. Another essential limitation is that these two conditions must be exactly equal in amount, not merely to produce the effects of attraction, but in every other case; for it is impossible so to arrange things that there shall be present or be evolved more electric power of the one kind than of the other. Another limitation is that they must be in physical relation to each other; and that when a positive and a negative electrified surface are thus associated, we cannot cut off this relation except by transferring the forces of these surfaces to equal amounts of the contrary forces provided elsewhere. Another limitation is that the power is definite in amount. If a ball *a* be charged with 10 of positive electricity, it may be made to act with that amount of power on another ball *b* charged with 10 of negative electricity; but if 5 of its power be taken up by a third ball *c* charged with negative electricity, then it can only act with 5 of power on ball *a*, and that ball must find or evolve 5 of positive power elsewhere: this is quite unlike what occurs with gravity, a power that presents us with nothing dual in its character. Finally, the electric force acts in curved lines. If a ball be electrified positively and insulated in the air, and a round metallic plate be placed about 12 or 15 inches off, facing it and uninsulated, the latter will be found, by the necessity mentioned above, in a negative condition; but it is not negative only on the side facing the ball, but on the other or outer face also, as may be shown by a carrier applied there, or by a strip of gold or silver leaf hung against that outer face. Now the power affecting this face does not pass through the uninsulated plate, for the thinnest gold leaf is able to stop the inductive action, but round the edges of the face, and therefore acts in curved lines. All these points indicate the existence of physical lines of electric force:—the absolutely essential relation of positive and negative surfaces to each other, and their dependence on each other contrasted with the known mobility of the forces, admit of no other conclusion. The action also in curved lines must depend upon a physical line of force. And there is a third important character of the force leading to the same result, namely its affection by media having different specific inductive capacities.

When we pass to Dynamic Electricity the evidence of physical lines of force is far more patent. A voltaic battery having its extremities connected by a conducting medium, has what has been expressively called a current of force running round the circuit, but this current is an axis of power having equal and contrary forces in opposite directions. It consists of lines of force which are compressed or expanded according to the transverse action of the conductor, which changes in direction with the form of the conductor, which are found in every part of the conductor, and can be taken out from any place by channels properly appointed for the purpose; and nobody doubts that they are physical lines of force.

Finally as regards a Magnet, which is the object of the present discourse. A magnet presents a system of forces perfect in itself, and able, therefore, to exist by its own mutual relations. It has the dual and antithetic character belonging to both static and dynamic electricity; and this is made manifest by what are called its polarities, *i. e.* by the opposite powers of like kind found at and towards its extremities. These powers are found to be absolutely equal to each other; one cannot be changed in any degree as to amount without an equal change of the other; and this is true when the opposite polarities of a magnet are not related to each other, but to the polarities of other magnets. The polarities, or the *northness* and *southness* of a magnet are not only related to each other, through or within the magnet itself, but they are also related externally to opposite polarities (in the manner of static electric induction), or they cannot exist; and this external relation involves and necessitates an exactly equal amount of the new opposite polarities to which those of the magnet are related. So that if the force of a magnet *a* is related to that of another magnet *b*, it cannot act on a third magnet *c* without being taken off from *b*, to an amount proportional to its action on *c*. The lines of magnetic force are shown by the moving wire to exist both within and outside of the magnet; also they are shown to be closed curves passing in one part of their course through the magnet; and the amount of those within the magnet at its equator is exactly equal in force to the amount in any section including the whole of those on the outside. The lines of force outside a magnet can be affected in their direction by the use of various media placed in their course. A magnet can in no way be procured having only one magnetism, or even the smallest excess of northness or southness one over the other. When the polarities of a magnet are not related externally to the forces of other magnets, then they are related to each other: *i. e.* the northness and southness of an isolated magnet are externally dependent on and sustained by each other.

Now all these facts, and many more, point to the existence of physical lines of force external to the magnets as well as within. They exist in curved as well as in straight lines; for if we conceive of an isolated straight bar-magnet, or more especially of a round disc of steel magnetized regularly, so that its magnetic axis shall be in one diameter, it is evident that the polarities must be related to each other externally by curved lines of force; for no straight line can at the same time touch two points having northness and southness. Curved lines of force can, as I think, only consist with physical lines of force.

The phænomena exhibited by the moving wire confirm the same conclusion. As the wire moves across the lines of force, a current of electricity passes or tends to pass through it, there being no such current before the wire is moved. The wire when quiescent has no such current, and when it moves it need not pass into places where the magnetic force is greater or less. It may travel in such a course that if a magnetic needle were carried through the same course it would be entirely unaffected magnetically, *i. e.* it would be a matter of absolute indifference to the needle whether it were moving or still. Matters may be so arranged that the wire when still shall have the same diamagnetic force as the medium surrounding the magnet, and so in no way cause disturbance of the lines of force passing through both; and yet when the wire moves, a current of electricity shall be generated in it. The mere fact of motion cannot have produced this current: there must have been a state or condition around the magnet and sustained by it, within the range of which the wire was placed; and this state shows the physical constitution of the lines of magnetic force.

What this state is, or upon what it depends, cannot as yet be declared. It may depend upon the æther, as a ray of light does, and an association has already been shown between light and magnetism. It may depend upon a state of tension, or a state of vibration, or perhaps some other state analogous to the electric current, to which the magnetic forces are so intimately related. Whether it of necessity requires matter for its sustentation will depend upon what is understood by the term matter. If that is to be confined to ponderable or gravitating substances, then matter is not essential to the physical lines of magnetic force any more than to a ray of light or heat; but if in the assumption of an æther we admit it to be a species of matter, then the lines of force may depend upon some function of it. Experimentally mere space is magnetic; but then the idea of such mere space must include that of the æther, when one is talking on that belief; or if hereafter any other conception of the state or condition of space rise up, it must be admitted into the view of that, which just now in relation to experiment is called mere space. On the other hand it is, I think, an ascertained fact, that ponderable matter is not essential to the existence of physical lines of magnetic force.

*2. James Clerk Maxwell*
*From* Treatise on Electricity and Magnetism,
*3rd Edition, Chapters VI and VII*

# Dynamical Theory of Electromagnetism

568.] WE have shewn, in Art. 552, that, when an electric current exists in a conducting circuit, it has a capacity for doing a certain amount of mechanical work, and this independently of any external electromotive force maintaining the current. Now capacity for performing work is nothing else than energy, in whatever way it arises, and all energy is the same in kind, however it may differ in form. The energy of an electric current is either of that form which consists in the actual motion of matter, or of that which consists in the capacity for being set in motion, arising from forces acting between bodies placed in certain positions relative to each other.

The first kind of energy, that of motion, is called Kinetic energy, and when once understood it appears so fundamental a fact of nature that we can hardly conceive the possibility of resolving it into anything else. The second kind of energy, that depending on position, is called Potential energy, and is due to the action of what we call forces, that is to say, tendencies towards change of relative position. With respect to these forces, though we may accept their existence as a demonstrated fact, yet we always feel that every explanation of the mechanism by which bodies are set in motion forms a real addition to our knowledge.

569.] The electric current cannot be conceived except as a kinetic phenomenon. Even Faraday, who constantly endeavoured to emancipate his mind from the influence of those suggestions which the words 'electric current' and 'electric fluid' are too apt to carry with them, speaks of the electric current as 'something progressive, and not a mere arrangement*.'

* *Exp. Res.*, 283.

The effects of the current, such as electrolysis, and the transfer of electrification from one body to another, are all progressive actions which require time for their accomplishment, and are therefore of the nature of motions.

As to the velocity of the current, we have shewn that we know nothing about it, it may be the tenth of an inch in an hour, or a hundred thousand miles in a second*. So far are we from knowing its absolute value in any case, that we do not even know whether what we call the positive direction is the actual direction of the motion or the reverse.

But all that we assume here is that the electric current involves motion of some kind. That which is the cause of electric currents has been called Electromotive Force. This name has long been used with great advantage, and has never led to any inconsistency in the language of science. Electromotive force is always to be understood to act on electricity only, not on the bodies in which the electricity resides. It is never to be confounded with ordinary mechanical force, which acts on bodies only, not on the electricity in them. If we ever come to know the formal relation between electricity and ordinary matter, we shall probably also know the relation between electromotive force and ordinary force.

570.] When ordinary force acts on a body, and when the body yields to the force, the work done by the force is measured by the product of the force into the amount by which the body yields. Thus, in the case of water forced through a pipe, the work done at any section is measured by the fluid pressure at the section multiplied into the quantity of water which crosses the section.

In the same way the work done by an electromotive force is measured by the product of the electromotive force into the quantity of electricity which crosses a section of the conductor under the action of the electromotive force.

The work done by an electromotive force is of exactly the same kind as the work done by an ordinary force, and both are measured by the same standards or units.

Part of the work done by an electromotive force acting on a

* *Exp. Res.*, 1648.

conducting circuit is spent in overcoming the resistance of the circuit, and this part of the work is thereby converted into heat. Another part of the work is spent in producing the electromagnetic phenomena observed by Ampère, in which conductors are made to move by electromagnetic forces. The rest of the work is spent in increasing the kinetic energy of the current, and the effects of this part of the action are shewn in the phenomena of the induction of currents observed by Faraday.

We therefore know enough about electric currents to recognise, in a system of material conductors carrying currents, a dynamical system which is the seat of energy, part of which may be kinetic and part potential.

The nature of the connexions of the parts of this system is unknown to us, but as we have dynamical methods of investigation which do not require a knowledge of the mechanism of the system, we shall apply them to this case.

We shall first examine the consequences of assuming the most general form for the function which expresses the kinetic energy of the system.

571.] Let the system consist of a number of conducting circuits, the form and position of which are determined by the values of a system of variables $x_1$, $x_2$, &c., the number of which is equal to the number of degrees of freedom of the system.

If the whole kinetic energy of the system were that due to the motion of these conductors, it would be expressed in the form

$$T = \tfrac{1}{2}(x_1 x_1)\dot{x}_1^2 + \&c. + (x_1 x_2)\dot{x}_1 \dot{x}_2 + \&c.,$$

where the symbols $(x_1 x_1)$, &c. denote the quantities which we have called moments of inertia, and $(x_1 x_2)$, &c. denote the products of inertia.

If $X'$ is the impressed force, tending to increase the coordinate $x$, which is required to produce the actual motion, then, by Lagrange's equation,

$$\frac{d}{dt}\frac{dT}{d\dot{x}} - \frac{dT}{dx} = X'.$$

When $T$ denotes the energy due to the visible motion only, we shall indicate it by the suffix $_m$, thus, $T_m$.

But in a system of conductors carrying electric currents, part of the kinetic energy is due to the existence of these currents. Let the motion of the electricity, and of anything whose motion is governed by that of the electricity, be determined by another set of coordinates $y_1$, $y_2$, &c., then $T$ will be a homogeneous function of squares and products of all the velocities of the two sets of coordinates. We may therefore divide $T$ into three portions, in the first of which, $T_m$, the velocities of the coordinates $x$ only occur, while in the second, $T_e$, the velocities of the coordinates $y$ only occur, and in the third, $T_{me}$, each term contains the product of the velocities of two coordinates of which one is an $x$ and the other a $y$.

We have therefore

$$T = T_m + T_e + T_{me},$$

where

$$T_m = \tfrac{1}{2}(x_1 x_1)\dot{x}_1^2 + \&c. + (x_1 x_2)\dot{x}_1 \dot{x}_2 + \&c.,$$

$$T_e \ \tfrac{1}{2}(y_1 y_1)\dot{y}_1^2 + \&c. + (y_1 y_2)\dot{y}_1 \dot{y}_2 + \&c.,$$

$$T_{me} = (x_1 y_1)\dot{x}_1 \dot{y}_1 + \&c.$$

572.] In the general dynamical theory, the coefficients of every term may be functions of all the coordinates, both $x$ and $y$. In the case of electric currents, however, it is easy to see that the coordinates of the class $y$ do not enter into the coefficients.

For, if all the electric currents are maintained constant, and the conductors at rest, the whole state of the field will remain constant. But in this case the coordinates $y$ are variable, though the velocities $\dot{y}$ are constant. Hence the coordinates $y$ cannot enter into the expression for $T$, or into any other expression of what actually takes place.

Besides this, in virtue of the equation of continuity, if the conductors are of the nature of linear circuits, only one variable is required to express the strength of the current in each conductor. Let the velocities $\dot{y}_1$, $\dot{y}_2$, &c. represent the strengths of the currents in the several conductors.

All this would be true, if, instead of electric currents, we had currents of an incompressible fluid running in flexible tubes. In this case the velocities of these currents would enter into the expression for $T$, but the coefficients would depend only on the variables $x$, which determine the form and position of the tubes.

In the case of the fluid, the motion of the fluid in one tube does not directly affect that of any other tube, or of the fluid in it. Hence, in the value of $T_e$, only the squares of the velocities $\dot{y}$, and not their products, occur, and in $T_{me}$ any velocity $\dot{y}$ is associated only with those velocities of the form $\dot{x}$ which belong to its own tube.

In the case of electrical currents we know that this restriction does not hold, for the currents in different circuits act on each other. Hence we must admit the existence of terms involving products of the form $\dot{y}_1 \dot{y}_2$, and this involves the existence of something in motion, whose motion depends on the strength of both electric currents $\dot{y}_1$ and $\dot{y}_2$. This moving matter, whatever it is, is not confined to the interior of the conductors carrying the two currents, but probably extends throughout the whole space surrounding them.

573.] Let us next consider the form which Lagrange's equations of motion assume in this case. Let $X'$ be the impressed force corresponding to the coordinate $x$, one of those which determine the form and position of the conducting circuits. This is a force in the ordinary sense, a tendency towards change of position. It is given by the equation

$$X' = \frac{d}{dt}\frac{dT}{d\dot{x}} - \frac{dT}{dx}.$$

We may consider this force as the sum of three parts, corresponding to the three parts into which we divided the kinetic energy of the system, and we may distinguish them by the same suffixes. Thus

$$X' = X'_m + X'_e + X'_{me}.$$

The part $X'_m$ is that which depends on ordinary dynamical considerations, and we need not attend to it.

Since $T_e$ does not contain $\dot{x}$, the first term of the expression for $X'_e$ is zero, and its value is reduced to

$$X'_e = -\frac{dT_e}{dx}.$$

This is the expression for the mechanical force which must be applied to a conductor to balance the electromagnetic force, and it asserts that it is measured by the rate of *diminution* of the purely electrokinetic energy due to the variation of the coordinate $x$. The electromagnetic force, $X_e$, which brings this external mechanical force into play, is equal and opposite to $X'_e$, and is therefore measured by the rate of *increase* of the electrokinetic energy corresponding to an increase of the coordinate $x$. The value of $X_e$, since it depends on squares and products of the currents, remains the same if we reverse the directions of all the currents.

The third part of $X'$ is

$$X'_{me} = \frac{d}{dt}\frac{dT_{me}}{d\dot{x}} - \frac{dT_{me}}{dx}.$$

The quantitity $T_{me}$ contains only products of the form $\dot{x}\dot{y}$, so that $\frac{dT_{me}}{d\dot{x}}$ is a linear function of the strengths of the currents $\dot{y}$. The first term, therefore, depends on the rate of variation of the strengths of the currents, and indicates a mechanical force on the conductor, which is zero when the currents are constant, and which is positive or negative according as the currents are increasing or decreasing in strength.

The second term depends, not on the variation of the currents, but on their actual strengths. As it is a linear function with respect to these currents, it changes sign when the currents change sign. Since every term involves a velocity $\dot{x}$, it is zero when the conductors are at rest. There are also terms arising from the time variations of the coefficients of $\dot{y}$ in $\frac{dT_{me}}{d\dot{x}}$: these remarks apply also to them.

We may therefore investigate these terms separately. If the

conductors are at rest, we have only the first term to deal with. If the currents are constant, we have only the second.

574.] As it is of great importance to determine whether any part of the kinetic energy is of the form $T_{me}$, consisting of products of ordinary velocities and strengths of electric currents, it is desirable that experiments should be made on this subject with great care.

The determination of the forces acting on bodies in rapid motion is difficult. Let us therefore attend to the first term, which depends on the variation of the strength of the current.

If any part of the kinetic energy depends on the product of an ordinary velocity and the strength of a current, it will probably be most easily observed when the velocity and the current are in the same or in opposite directions. We therefore take a circular coil of a great many windings, and suspend it by a fine vertical wire, so that its windings are horizontal, and the coil is capable of rotating about a vertical axis, either in the same direction as the current in the coil, or in the opposite direction.

We shall suppose the current to be conveyed into the coil by means of the suspending wire, and, after passing round the windings, to complete its circuit by passing downwards through a wire in the same line with the suspending wire and dipping into a cup of mercury.

Since the action of the horizontal component of terrestrial magnetism would tend to turn this coil round a horizontal axis when the current flows through it, we shall suppose that the horizontal component of terrestrial magnetism is exactly neutralized by means of fixed magnets, or that the experiment is made at the magnetic pole. A vertical mirror is attached to the coil to detect any motion in azimuth.

Now let a current be made to pass through the coil in the direction N.E.S.W. If electricity were a fluid like water, flowing along the wire, than, at the moment of starting the current, and as long as its velocity is increasing, a force would require to be supplied to produce the angular momentum of the fluid in passing round the coil, and as this must be supplied by the elasticity of the suspending wire, the coil would at first rotate in the opposite direction or W.S.E.N., and this would be detected by means of the mirror. On stopping the current there would be another movement of the mirror, this time in the same direction as that of the current.

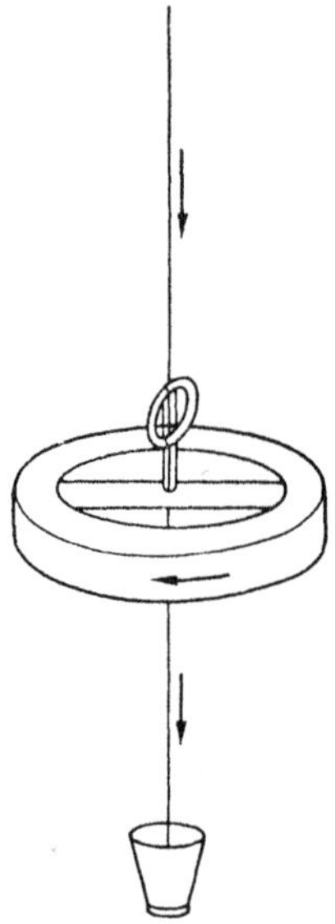

FIG. 33.

No phenomenon of this kind has yet been observed. Such an action, if it existed, might be easily distinguished from the already known actions of the current by the following peculiarities.

(1) It would occur only when the strength of the current varies, as when contact is made or broken, and not when the current is constant.

All the known *mechanical* actions of the current depend on the strength of the currents, and not on the rate of variation. The electromotive action in the case of induced currents cannot be confounded with this electromagnetic action.

(2) The direction of this action would be reversed when that of all the currents in the field is reversed.

All the known mechanical actions of the current remain the same when all the currents are reversed, since they depend on squares and products of these currents.

If any action of this kind were discovered, we should be able to regard one of the so-called kinds of electricity, either the positive or the negative kind, as a real substance, and we should be able to describe the electric current as a true motion of this substance in a particular direction. In fact, if electrical motions were in any way comparable with the motions of ordinary matter, terms of the form $T_{me}$ would exist, and their existence would be manifested by the mechanical force $X_{me}$.

According to Fechner's hypothesis, that an electric current consists of two equal currents of positive and negative electricity, flowing in opposite directions through the same conductor, the terms of the second class $T_{me}$ would vanish, each term belonging to the positive current being accompanied by an equal term of opposite sign belonging to the negative current, and the phenomena depending on these terms would have no existence.

It appears to me, however, that while we derive great advantage from the recognition of the many analogies between the electric current and a current of material fluid, we must carefully avoid making any assumption not warranted by experimental evidence, and that there is, as yet, no experimental evidence to shew whether the electric current is really a current of a material substance, or a double current, or whether its velocity is great or small as measured in feet per second.

A knowledge of these things would amount to at least the beginnings of a complete dynamical theory of electricity, in which we should regard electrical action, not, as in this treatise, as a phenomenon due to an unknown cause, subject only to the general laws of dynamics, but as the result of known motions of known portions of matter, in which not only the total effects and final results, but the whole intermediate mechanism and details of the motion, are taken as the objects of study.

575.] The experimental investigation of the second term of $X_{me}$, namely $\frac{dT_{me}}{dx}$, is more difficult, as it involves the observation of the effect of forces on a body in rapid motion.

The apparatus shewn in Fig. 34, which I had constructed in 1861, is intended to test the existence of a force of this kind.

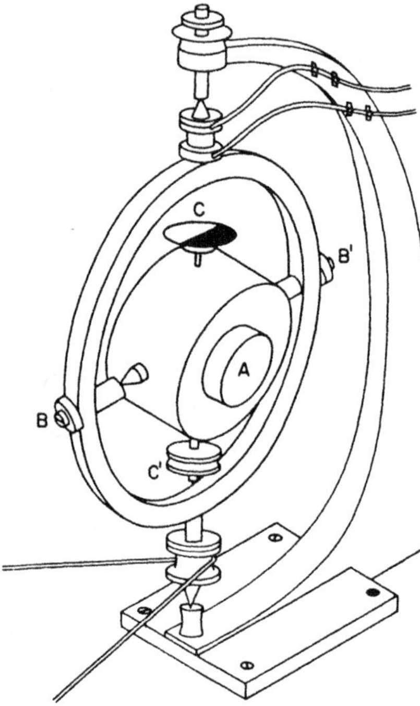


FIG. 34.

The electromagnet $A$ is capable of rotating about the horizontal axis $BB'$, within a ring which itself revolves about a vertical axis.

Let $A$, $B$, $C$ be the moments of inertia of the electromagnet about the axis of the coil, the horizontal axis $BB'$, and a third axis $CC'$ respectively.

Let $\theta$ be the angle which $CC'$ makes with the vertical, $\phi$ the azimuth of the axis $BB'$, and $\psi$ a variable on which the motion of electricity in the coil depends.

Then the kinetic energy $T$ of the electromagnet may be written

$$2T = A\dot{\phi}^2 \sin^2\theta + B\dot{\theta}^2 + C\dot{\phi}^2 \cos^2\theta + E(\dot{\phi}\sin\theta + \dot{\psi})^2,$$

where $E$ is a quantity which may be called the moment of inertia of the electricity in the coil.

If $\Theta$ is the moment of the impressed force tending to increase $\theta$, we have, by the equations of dynamics,

$$\Theta = B\frac{d^2\theta}{dt^2} - \{(A - C)\dot{\phi}^2 \sin\theta\cos\theta + E\dot{\phi}\cos\theta(\dot{\phi}\sin\theta + \dot{\psi})\}$$

By making $\Psi$, the impressed force tending to increase $\psi$, equal to zero, we obtain

$$\dot{\phi}\sin\theta + \dot{\psi} = \gamma,$$

a constant, which we may consider as representing the strength of the current in the coil.

If $C$ is somewhat greater than $A$; $\Theta$ will be zero, and the equilibrium about the axis $BB'$ will be stable when

$$\sin\theta = \frac{E\gamma}{(C - A)\dot{\phi}}.$$

This value of $\theta$ depends on that of $\gamma$, the electric current, and is positive or negative according to the direction of the current.

The current is passed through the coil by its bearings at $B$ and $B'$, which are connected with the battery by means of springs rubbing on metal rings placed on the vertical axis.

To determine the value of $\theta$, a disk of paper is placed at $C$, divided by a diameter parallel to $BB'$ into two parts, one of which is painted red and the other green.

When the instrument is in motion a red circle is seen at $C$ when $\theta$ is positive, the radius of which indicates roughly the value of $\theta$. When $\theta$ is negative, a green circle is seen at $C$.

By means of nuts working on screws attached to the electromagnet, the axis $CC'$ is adjusted to be a principal axis having its moment of inertia just exceeding that round the axis $A$, so as to make the instrument very sensitive to the action of the force if it exists.

The chief difficulty in the experiments arose from the disturbing action of the earth's magnetic force, which caused the electromagnet to act like a dip-needle. The results obtained were on this account very rough, but no evidence of any change in $\theta$ could be obtained even when an iron core was inserted in the coil, so as to make it a powerful electromagnet.

If, therefore, a magnet contains matter in rapid rotation, the angular momentum of this rotation must be very small compared with any quantities which we can measure, and we have as yet no evidence of the existence of the terms $T_{me}$ derived from their mechanical action.

576.] Let us next consider the forces acting on the currents of electricity, that is, the electromotive forces.

Let $Y$ be the effective electromotive force due to induction, the electromotive force which must act on the circuit from without to balance it is $Y' = -Y$, and, by Lagrange's equation,

$$Y = -Y' = -\frac{d}{dt}\frac{dT}{d\dot{y}} + \frac{dT}{dy}.$$

Since there are no terms in $T$ involving the coordinate $y$, the second term is zero, and $Y$ is reduced to its first term. Hence, electromotive force cannot exist in a system at rest, and with constant currents.

Again, if we divide $Y$ into three parts $Y_m$, $Y_e$, and $Y_{me}$, corresponding to the three parts of $T$, we find that, since $T_m$ does not contain $\dot{y}$, $Y_m = 0$.

We also find

$$Y_e = -\frac{d}{dt}\frac{dT_e}{d\dot{y}}.$$

Here $\frac{dT_e}{d\dot{y}}$ is a linear function of the currents, and this part of the electromotive force is equal to the rate of change of this function.

This is the electromotive force of induction discovered by Faraday. We shall consider it more at length afterwards.

577.] From the part of $T$, depending on velocities multiplied by currents, we find

$$Y_{me} = -\frac{d}{dt}\frac{dT_{me}}{d\dot{y}}.$$

Now $\frac{dT_{me}}{d\dot{y}}$ is a linear function of the velocities of the conductors. If, therefore, any terms of $T_{me}$ have an actual existence, it would be possible to produce an electromotive force independently of all existing currents by simply altering the velocities of the conductors. For instance, in the case of the suspended coil at Art. 574, if, when the coil is at rest, we suddenly set it in rotation about the vertical axis, an electromotive force would be called into action proportional to the acceleration of this motion. It would vanish when the motion became uniform, and be reversed when the motion was retarded.

Now few scientific observations can be made with greater precision than that which determines the existence or non-existence of a current by means of a galvanometer. The delicacy of this method far exceeds that of most of the arrangements for measuring the mechanical force acting on a body. If, therefore, any currents could be produced in this way they would be detected, even if they were very feeble. They would be distinguished from ordinary currents of induction by the following characteristics.

(1) They would depend entirely on the motions of the conductors, and in no degree on the strength of currents or magnetic forces already in the field.

(2) They would depend not on the absolute velocities of the conductors, but on their accelerations, and on squares and products of velocities, and they would change when the acceleration becomes a retardation, though the absolute velocity is the same.

Now in all the cases actually observed, the induced currents depend altogether on the strength and the variation of currents in the field, and cannot be excited in a field devoid of magnetic force and of currents. In so far as they depend on the motion of conductors, they depend on the absolute velocity, and not on the change of velocity of these motions.

We have thus three methods of detecting the existence of the terms of the form $T_{me}$, none of which have hitherto led to any positive result. I have pointed them out with the greater care because it appears to me important that we should attain the greatest amount of certitude within our reach on a point bearing so strongly on the true theory of electricity.

Since, however, no evidence has yet been obtained of such terms, I shall now proceed on the assumption that they do not exist, or at least that they produce no sensible effect, an assumption which will considerably simplify our dynamical theory. We shall have occasion, however, in discussing the relation of magnetism to light, to shew that the motion which constitutes light may enter as a factor into terms involving the motion which constitutes magnetism.

# Theory of Electric Circuits

578.] WE may now confine our attention to that part of the kinetic energy of the system which depends on squares and products of the strengths of the electric currents. We may call this the Electrokinetic Energy of the system. The part depending on the motion of the conductors belongs to ordinary dynamics, and we have seen that the part depending on products of velocities and currents does not exist.

Let $A_1$, $A_2$, &c. denote the different conducting circuits. Let their form and relative position be expressed in terms of the variables $x_1$, $x_2$, &c. the number of which is equal to the number of degrees of freedom of the mechanical system. We shall call these the Geometrical Variables.

Let $y_1$ denote the quantity of electricity which has crossed a given section of the conductor $A_1$ since the beginning of the time $t$. The strength of the current will be denoted by $\dot{y}_1$, the fluxion of this quantity.

We shall call $\dot{y}_1$ the actual current, and $y_1$ the integral current. There is one variable of this kind for each circuit in the system.

Let $T$ denote the electrokinetic energy of the system. It is a homogeneous function of the second degree with respect to the strengths of the currents, and is of the form

$$T = \tfrac{1}{2}L_1\dot{y}_1^2 + \tfrac{1}{2}L_2\dot{y}_2^2 + \text{\&c.} + M_{12}\dot{y}_1\dot{y}_2 + \text{\&c.}, \qquad (1)$$

where the coefficients $L$, $M$, &c. are functions of the geometrical variables $x_1$, $x_2$, &c. The electrical variables $y_1$, $y_2$ do not enter into the expression.

We may call $L_1$, $L_2$, &c. the electric moments of inertia of the circuits $A_1$, $A_2$, &c., and $M_{12}$ the electric product of inertia of the two circuits $A_1$ and $A_2$. When we wish to avoid the language of the dynamical theory, we shall call $L_1$ the coefficient of self-induction of the circuit $A_1$, and $M_{12}$ the coefficient of mutual induction of the circuits $A_1$ and $A_2$. $M_{12}$ is also called the potential of the circuit $A_1$ with respect to $A_2$. These quantities depend only on the form and relative position of the circuits. We shall find that in the electromagnetic system of measurement they are quantities of the dimension of a line. See Art. 627.

By differentiating $T$ with respect to $\dot{y}_1$ we obtain the quantity $p_1$, which, in the dynamical theory, may be called the momentum corresponding to $y_1$. In the electric theory we shall call $p_1$ the electrokinetic momentum of the circuit $A_1$. Its value is

$$p_1 = L_1\dot{y}_1 + M_{12}\dot{y}_2 + \&c.$$

The electrokinetic momentum of the circuit $A_1$ is therefore made up of the product of its own current into its coefficient of self-induction, together with the sum of the products of the currents in the other circuits, each into the coefficient of mutual induction of $A_1$ and that other circuit.

## Electromotive Force

579.] Let $E$ be the impressed electromotive force in the circuit $A$, arising from some cause, such as a voltaic or thermo-electric battery, which would produce a current independently of magneto-electric induction.

Let $R$ be the resistance of the circuit, then, by Ohm's law, an electromotive force $R\dot{y}$ is required to overcome the resistance, leaving an electromotive force $E - R\dot{y}$ available for changing the momentum of the circuit. Calling this force $Y'$, we have, by the general equations,

$$Y' = \frac{dp}{dt} - \frac{dT}{dy},$$

but since $T$ does not involve $y$, the last term disappears.

Hence, the equation of electromotive force is

$$E - R\dot{y} = Y' = \frac{dp}{dt},$$

or

$$E = R\dot{y} + \frac{dp}{dt}.$$

The impressed electromotive force $E$ is therefore the sum of two parts. The first, $R\dot{y}$, is required to maintain the current $\dot{y}$ against the resistance $R$. The second part is required to increase the electromagnetic momentum $p$. This is the electromotive force which must be supplied from sources independent of magneto-electric induction. The electromotive-force arising from magneto-electric induction alone is evidently $-\frac{dp}{dt}$, or, *the rate of decrease of the electrokinetic momentum of the circuit.*

## Electromagnetic Force

580.] Let $X'$ be the impressed mechanical force arising from external causes, and tending to increase the variable $x$. By the general equations

$$X' = \frac{d}{dt}\frac{dT}{d\dot{x}} - \frac{dT}{dx}.$$

Since the expression for the electrokinetic energy does not contain the velocity ($\dot{x}$), the first term of the second member disappears, and we find

$$X' = -\frac{dT}{dx}.$$

Here $X'$ is the external force required to balance the forces arising from electrical causes. It is usual to consider this force as the reaction against the electromagnetic force, which we shall call $X$, and which is equal and opposite to $X'$.

Hence

$$X = \frac{dT}{dx},$$

or, *the electromagnetic force tending to increase any variable is equal to the rate of increase of the electrokinetic energy per unit increase of that variable, the currents being maintained constant.*

If the currents are maintained constant by a battery during a displacement in which a quantity, $W$, of work is done by electromotive force, the electrokinetic energy of the system will be at the same time increased by $W$. Hence the battery will be drawn upon for a double quantity of energy, or $2W$, in addition to that which is spent in generating heat in the circuit. This was first pointed out by Sir W. Thomson*. Compare this result with the electrostatic property in Art. 93.

## Case of Two Circuits

581.] Let $A_1$ be called the Primary Circuit, and $A_2$ the Secondary Circuit. The electrokinetic energy of the system may be written

$$T = \tfrac{1}{2} L \dot{y}_1^2 + M \dot{y}_1 \dot{y}_2 + \tfrac{1}{2} N \dot{y}_2^2,$$

where $L$ and $N$ are the coefficients of self-induction of the primary and secondary circuits respectively, and $M$ is the coefficient of their mutual induction.

Let us suppose that no electromotive force acts on the secondary circuit except that due to the induction of the primary current. We have then

$$E_2 = R_2 \dot{y}_2 + \frac{d}{dt}(M \dot{y}_1 + N \dot{y}_2) = 0.$$

Integrating this equation with respect to $t$, we have

$$R_2 y_2 + M \dot{y}_1 + N \dot{y}_2 = C, \text{ a constant,}$$

where $y_2$ is the integral current in the secondary circuit.

The method of measuring an integral current of short duration will be described in Art. 748, and it is easy in most cases to ensure that the duration of the secondary current shall be very short.

Let the values of the variable quantities in the equation at the end of the time $t$ be accented, then, if $y_2$ is the integral current,

* Nichol's *Cyclopaedia of the Physical Sciences*, ed. 1860, article 'Magnetism, Dynamical Relations of.'

or the whole quantity of electricity which flows through a section of the secondary circuit during the time $t$,

$$R_2 y_2 = M\dot{y}_1 + N\dot{y}_2 - (M'\dot{y}_1' + N'\dot{y}_2').$$

If the secondary current arises entirely from induction, its initial value $\dot{y}_2$ must be zero if the primary current is constant and the conductors are at rest before the beginning of the time $t$.

If the time $t$ is sufficient to allow the secondary current to die away, $\dot{y}_2'$, its final value, is also zero, so that the equation becomes

$$R_2 y_2 = M\dot{y}_1 - M'\dot{y}_1.$$

The integral current of the secondary circuit depends in this case on the initial and final values of $M\dot{y}_1$.

## Induced Currents

582.] Let us begin by supposing the primary circuit broken, or $\dot{y}_1 = 0$, and let a current $\dot{y}_1'$ be established in it when contact is made.

The equation which determines the secondary integral current is

$$R_2 y_2 = -M_1''\dot{y}.$$

When the circuits are placed side by side, and in the same direction, $M'$ is a positive quantity. Hence, when contact is made in the primary circuit, a negative current is induced in the secondary circuit.

When the contact is broken in the primary circuit, the primary current ceases, and the induced integral current is $y_2$, where

$$R_2 y_2 = M\dot{y}_1.$$

The secondary current is in this case positive.

If the primary current is maintained constant, and the form or relative position of the circuits altered so that $M$ becomes $M'$, the integral secondary current is $y_2$, where

$$R_2 y_2 = (M - M')\dot{y}_1.$$

In the case of two circuits placed side by side and in the same direction $M$ diminishes as the distance between the circuits increases. Hence, the induced current is positive when this distance is increased and negative when it is diminished.

These are the elementary cases of induced currents described in Art. 530.

### Mechanical Action between the Two Circuits

583.] Let $x$ be any one of the geometrical variables on which the form and relative position of the circuits depend, the electromagnetic force tending to increase $x$ is

$$X = \tfrac{1}{2}\dot{y}_1^2 \frac{dL}{dx} + \dot{y}_1\dot{y}_2 \frac{dM}{dx} + \tfrac{1}{2}\dot{y}_2^2 \frac{dN}{dx}.$$

If the motion of the system corresponding to the variation of $x$ is such that each circuit moves as a rigid body, $L$ and $N$ will be independent of $x$, and the equation will be reduced to the form

$$X = \dot{y}_1\dot{y}_2 \frac{dM}{dx}.$$

Hence, if the primary and secondary currents are of the same sign, the force $X$, which acts between the circuits, will tend to move them so as to increase $M$.

If the circuits are placed side by side, and the currents flow in the same direction. $M$ will be increased by their being brought nearer together. Hence the force $X$ is in this case an attraction.

584.] The whole of the phenomena of the mutual action of two circuits, whether the induction of currents or the mechanical force between them, depend on the quantity $M$, which we have called the coefficient of mutual induction. The method of calculating this quantity from the geometrical relations of the circuits is given in Art. 524, but in the investigations of the next chapter we shall not assume a knowledge of the mathematical form of this quantity. We shall consider it as deduced from experiments on induction, as, for instance, by observing the integral current when the secondary circuit is suddenly moved from a given position to an infinite distance, or to any position in which we know that $M = 0$.

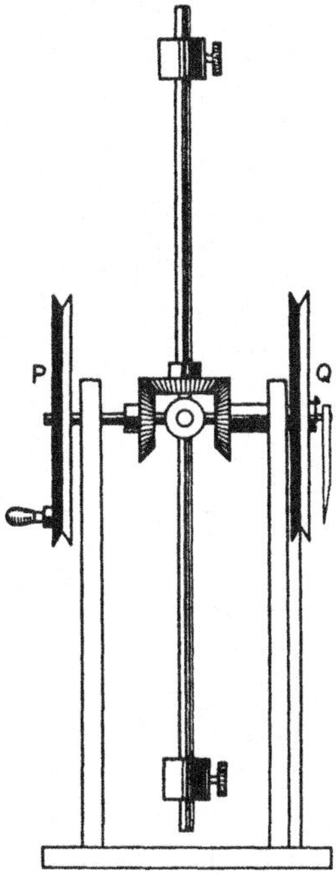


FIG. 34 *a*.

NOTE.*—{There is a model in the Cavendish Laboratory designed by Maxwell which illustrates very clearly the laws of the induction of currents. It is represented in Fig. 34 *a*. *P* and *Q* are two disks, the rotation of *P* represents the primary current, that of *Q* the secondary. These disks are connected together by a differential gearing. The intermediate wheel carries a fly-wheel the moment of inertia of which can be altered by moving weights inwards or outwards. The resistance of the secondary circuit is represented by the friction of a string passing over *Q* and kept tight by an elastic band. If the disk *P* is set in rotation (a current started in the primary) the disk *Q* will turn in the opposite direction (inverse current when the primary is started). When the velocity of rotation of *P* becomes uniform, *Q* is at rest (no current in the secondary when the primary current is constant); if the disk *P* is stopped, *Q* commences to rotate in the direction in which *P* was previously moving (direct current in the secondary on breaking the circuit). The effect of an iron core in increasing the induction can be illustrated by increasing the moment of inertia of the fly-wheel.}

* Added by J. J. Thomson for the third edition of the *Treatise*.

*XXV. A Dynamical Theory of the Electromagnetic Field* *

(Received October 27,—Read December 8, 1864)

PART I

# Introductory

(1) THE most obvious mechanical phenomenon in electrical and magnetical experiments is the mutual action by which bodies in certain states set each other in motion while still at a sensible distance from each other. The first step, therefore, in reducing these phenomena into scientific form, is to ascertain the magnitude and direction of the force acting between the bodies, and when it is found that this force depends in a certain way upon the relative position of the bodies and on their electric or magnetic condition, it seems at first sight natural to explain the facts by assuming the existence of something either at rest or in motion in each body, constituting its electric or magnetic state, and capable of acting at a distance according to mathematical laws.

In this way mathematical theories of statical electricity, of magnetism, of the mechanical action between conductors carrying currents, and of the induction of currents have been formed. In these theories the force acting between the two bodies is treated with reference only to the condition of the bodies and their relative position, and without any express consideration of the surrounding medium.

These theories assume, more or less explicitly, the existence of substances the particles of which have the property of acting on one another at a distance by attraction or repulsion. The most complete development of a theory of this kind is that of M. W.

* From the *Royal Society Transactions*, Vol. CLV.

Weber*, who has made the same theory include electrostatic and electromagnetic phenomena.

In doing so, however, he has found it necessary to assume that the force between two electric particles depends on their relative velocity, as well as on their distance.

This theory, as developed by MM. W. Weber and C. Neumann†, is exceedingly ingenious, and wonderfully comprehensive in its application to the phenomena of statical electricity, electromagnetic attractions, induction of currents and diamagnetic phenomena; and it comes to us with the more authority, as it has served to guide the speculations of one who has made so great an advance in the practical part of electric science, both by introducing a consistent system of units in electrical measurement, and by actually determining electrical quantities with an accuracy hitherto unknown.

(2) The mechanical difficulties, however, which are involved in the assumption of particles acting at a distance with forces which depend on their velocities are such as to prevent me from considering this theory as an ultimate one, though it may have been, and may yet be useful in leading to the coordination of phenomena.

I have therefore preferred to seek an explanation of the fact in another direction, by supposing them to be produced by actions which go on in the surrounding medium as well as in the excited bodies, and endeavouring to explain the action between distant bodies without assuming the existence of forces capable of acting directly at sensible distances.

(3) The theory I propose may therefore be called a theory of the *Electromagnetic Field*, because it has to do with the space in the neighbourhood of the electric or magnetic bodies, and it may be called a *Dynamical* Theory, because it assumes that in that space there is matter in motion, by which the observed electromagnetic phenomena are produced.

* "Electrodynamische Maassbestimmungen." *Leipzic Trans*. Vol. I. 1849, and Taylor's *Scientific Memoirs*, Vol. V. art. xiv.

† *Explicare tentatur quomodo fiat ut lucis planum polarizationis per vires electricas vel magnetica declinetur*.—Halis Saxonum, 1858.

(4) The electromagnetic field is that part of space which contains and surrounds bodies in electric or magnetic conditions.

It may be filled with any kind of matter, or we may endeavour to render it empty of all gross matter, as in the case of Geissler's tubes and other so-called vacua.

There is always, however, enough of matter left to receive and transmit the undulations of light and heat, and it is because the transmission of these radiations is not greatly altered when transparent bodies of measurable density are substituted for the so-called vacuum, that we are obliged to admit that the undulations are those of an aethereal substance, and not of the gross matter, the presence of which merely modifies in some way the motion of the aether.

We have therefore some reason to believe, from the phenomena of light and heat, that there is an aethereal medium filling space and permeating bodies, capable of being set in motion and of transmitting that motion from one part to another, and of communicating that motion to gross matter so as to heat it and affect it in various ways.

(5) Now the energy communicated to the body in heating it must have formerly existed in the moving medium, for the undulations had left the source of heat some time before they reached the body, and during that time the energy must have been half in the form of motion of the medium and half in the form of elastic resilience. From these considerations Professor W. Thomson has argued*, that the medium must have a density capable of comparison with that of gross matter, and has even assigned an inferior limit to that density.

(6) We may therefore receive, as a datum derived from a branch of science independent of that with which we have to deal, the existence of a pervading medium, of small but real density, capable of being set in motion, and of transmitting motion from one part to another with great, but not infinite, velocity.

* "On the Possible Density of the Luminiferous Medium, and on the Mechanical Value of a Cubic Mile of Sunlight," *Transactions of the Royal Society of Edinburgh* (1854), p. 57.

Hence the parts of this medium must be so connected that the motion of one part depends in some way on the motion of the rest; and at the same time these connexions must be capable of a certain kind of elastic yielding, since the communication of motion is not instantaneous, but occupies time.

The medium is therefore capable of receiving and storing up two kinds of energy, namely, the "actual" energy depending on the motions of its parts, and "potential" energy, consisting of the work which the medium will do in recovering from displacement in virtue of its elasticity.

The propagation of undulations consists in the continual transformation of one of these forms of energy into the other alternately, and at any instant the amount of energy in the whole medium is equally divided, so that half is energy of motion, and half is elastic resilience.

(7) A medium having such a constitution may be capable of other kinds of motion and displacement than those which produce the phenomena of light and heat, and some of these may be of such a kind that they may be evidenced to our senses by the phenomena they produce.

(8) Now we know that the luminiferous medium is in certain cases acted on by magnetism; for Faraday* discovered that when a plane polarized ray traverses a transparent diamagnetic medium in the direction of the lines of magnetic force produced by magnets or currents in the neighbourhood, the plane of polarization is caused to rotate.

This rotation is always in the direction in which positive electricity must be carried round the diamagnetic body in order to produce the actual magnetization of the field.

M. Verdet† has since discovered that if a paramagnetic body, such as solution of perchloride of iron in ether, be substituted for the diamagnetic body, the rotation is in the opposite direction.

* *Experimental Researches*, Series XIX.

† *Comptes Rendus* (1856, second half year, p. 529, and 1857, first half year, p. 1209).

Now Professor W. Thomson* has pointed out that no distribution of forces acting between the parts of a medium whose only motion is that of the luminous vibrations, is sufficient to account for the phenomena, but that we must admit the existence of a motion in the medium depending on the magnetization, in addition to the vibratory motion which constitutes light.

It is true that the rotation by magnetism of the plane of polarization has been observed only in media of considerable density; but the properties of the magnetic field are not so much altered by the substitution of one medium for another, or for a vacuum, as to allow us to suppose that the dense medium does anything more than merely modify the motion of the ether. We have therefore warrantable grounds for inquiring whether there may not be a motion of the ethereal medium going on wherever magnetic effects are observed, and we have some reason to suppose that this motion is one of rotation, having the direction of the magnetic force as its axis.

(9) We may now consider another phenomenon observed in the electromagnetic field. When a body is moved across the lines of magnetic force it experiences what is called an electromotive force; the two extremities of the body tend to become oppositely electrified, and an electric current tends to flow through the body. When the electromotive force is sufficiently powerful, and is made to act on certain compound bodies, it decomposes them, and causes one of their components to pass towards one extremity of the body, and the other in the opposite direction.

Here we have evidence of a force causing an electric current in spite of resistance; electrifying the extremities of a body in opposite ways, a condition which is sustained only by the action of the electromotive force, and which, as soon as that force is removed, tends, with an equal and opposite force, to produce a counter current through the body and to restore the original electrical state of the body; and finally, if strong enough, tearing to pieces chemical compounds and carrying their components in opposite directions, while their natural tendency is to combine, and to

* *Proceedings of the Royal Society*, June 1856 and June 1861.

combine with a force which can generate an electromotive force in the reverse direction.

This, then, is a force acting on a body caused by its motion through the electromagnetic field, or by changes occurring in that field itself; and the effect of the force is either to produce a current and heat the body, or to decompose the body, or, when it can do neither, to put the body in a state of electric polarization,—a state of constraint in which opposite extremities are oppositely electrified, and from which the body tends to relieve itself as soon as the disturbing force is removed.

(10) According to the theory which I propose to explain, this "electromotive force" is the force called into play during the communication of motion from one part of the medium to another, and it is by means of this force that the motion of one part causes motion in another part. When electromotive force acts on a conducting circuit, it produces a current, which, as it meets with resistance, occasions a continual transformation of electrical energy into heat, which is incapable of being restored again to the form of electrical energy by any reversal of the process.

(11) But when electromotive force acts on a dielectric it produces a state of polarization of its parts similar in distribution to the polarity of the particles of a mass of iron under the influence of a magnet, and like the magnetic polarization, capable of being described as a state in which every particle has its opposite poles in opposite conditions*.

In a dielectric under the action of electromotive force, we may conceive that the electricity in each molecule is so displaced that one side is rendered positively and the other negatively electrical, but that the electricity remains entirely connected with the molecule and does not pass from one molecule to another. The effect of this action on the whole dielectric mass is to produce a general displacement of electricity in a certain direction. This displacement does not amount to a current, because when it has attained to a a certain value it remains constant, but it is the commencement

* Faraday, *Experimental Researches*, Series XI.; Mossotti, *Mem. della Soc. Italiana* (Modena), Vol. XXIV. Part 2, p. 49.

of a current, and its variations constitute currents in the positive or the negative direction according as the displacement is increasing or decreasing. In the interior of the dielectric there is no indication of electrification, because the electrification of the surface of any molecule is neutralized by the opposite electrification of the surface of the molecules in contact with it; but at the bounding surface of the dielectric where the electrification is not neutralized, we find the phenomena which indicate positive or negative electrification.

The relation between the electromotive force and the amount of electric displacement it produces depends on the nature of the dielectric, the same electromotive force producing generally a greater electric displacement in solid dielectrics, such as glass or sulphur, than in air.

(12) Here, then, we perceive another effect of electromotive force, namely electric displacement, which according to our theory is a kind of elastic yielding to the action of the force, similar to that which takes place in structures and machines owing to the want of perfect rigidity of the connexions.

(13) The practical investigation of the inductive capacity of dielectrics is rendered difficult on account of two disturbing phenomena. The first is the conductivity of the dielectric, which, though in many cases exceedingly small is not altogether insensible. The second is the phenomenon called electric absorption*, in virtue of which, when the dielectric is exposed to electromotive force, the electric displacement gradually increases, and when the electromotive force is removed, the dielectric does not instantly return to its primitive state, but only discharges a portion of its electrification, and when left to itself gradually acquires electrification on its surface, as the interior gradually becomes depolarized. Almost all solid dielectrics exhibit this phenomenon, which gives rise to the residual charge in the Leyden jar, and to several phenomena of electric cables described by Mr. F. Jenkin†.

* Faraday, *Experimental Researches*, 1233–1250.

† *Reports of British Association*, 1859, p. 248; and *Report of Committee of Board of Trade on Submarine Cables*, pp. 136 & 464.

(14) We have here two other kinds of yielding besides the yielding of the perfect dielectric, which we have compared to a perfectly elastic body. The yielding due to conductivity may be compared to that of a viscous fluid (that is to say, a fluid having great internal friction), or a soft solid on which the smallest force produces a permanent alteration of figure increasing with the time during which the force acts. The yielding due to electric absorption may be compared to that of a cellular elastic body containing a thick fluid in its cavities. Such a body, when subjected to pressure, is compressed by degrees on account of the gradual yielding of the thick fluid; and when the pressure is removed it does not at once recover its figure, because the elasticity of the substance of the body has gradually to overcome the tenacity of the fluid before it can regain complete equilibrium.

Several solid bodies in which no such structure as we have supposed can be found, seem to possess a mechanical property of this kind*; and it seems probable that the same substances, if dielectrics, may possess the analogous electrical property, and if magnetic, may have corresponding properties relating to the acquisition, retention, and loss of magnetic polarity.

(15) It appears therefore that certain phenomena in electricity and magnetism lead to the same conclusion as those of optics, namely, that there is an aethereal medium pervading all bodies, and modified only in degree by their presence; that the parts of this medium are capable of being set in motion by electric currents and magnets; that this motion is communicated from one part of the medium to another by forces arising from the connexions of those parts; that under the action of these forces there is a certain yielding depending on the elasticity of these connexions; and that therefore energy in two different forms may exist in the medium, the one form being the actual energy of motion of its parts, and the other being the potential energy stored up in the connexions, in virtue of their elasticity.

* As, for instance, the composition of glue, treacle, &c., of which small plastic figures are made, which after being distorted gradually recover their shape.

(16) Thus, then, we are led to the conception of a complicated mechanism capable of a vast variety of motion, but at the same time so connected that the motion of one part depends, according to definite relations, on the motion of other parts, these motions being communicated by forces arising from the relative displacement of the connected parts, in virture of their elasticity. Such a mechanism must be subject to the general laws of Dynamics, and we ought to be able to work out all the consequences of its motion, provided we know the form of the relation between the motions of the parts.

(17) We know that when an electric current is established in a conducting circuit, the neighbouring part of the field is characterized by certain magnetic properties, and that if two circuits are in the field, the magnetic properties of the field due to the two currents are combined. Thus each part of the field is in connexion with both currents, and the two currents are put in connexion with each other in virtue of their connexion with the magnetization of the field. The first result of this connexion that I propose to examine, is the induction of one current by another, and by the motion of conductors in the field.

The second result, which is deduced from this, is the mechanical action between conductors carrying currents. The phenomenon of the induction of currents has been deduced from their mechanical action by Helmholtz* and Thomson†. I have followed the reverse order, and deduced the mechanical action from the laws of induction. I have then described experimental methods of determining the quantities L, M, N, on which these phenomena depend.

(18) I then apply the phenomena of induction and attraction of currents to the exploration of the electromagnetic field, and the laying down systems of lines of magnetic force which indicate its magnetic properties. By exploring the same field with a magnet

* "Conservation of Force," *Physical Society of Berlin*, 1847; and Taylors' *Scientific Memoirs*, 1853, p. 114.

† *Reports of the British Association*, 1848; *Philosophical Magazine*, Dec. 1851.

I shew the distribution of its equipotential magnetic surfaces, cutting the lines of force at right angles.

In order to bring these results within the power of symbolical calculation, I then express them in the form of the General Equations of the Electromagnetic Field. These equations express—

(A) The relation between electric displacement, true conduction, and the total current, compounded of both.

(B) The relation between the lines of magnetic force and the inductive coefficients of a circuit, as already deduced from the laws of induction.

(C) The relation between the strength of a current and its magnetic effects, according to the electromagnetic system of measurement.

(D) The value of the electromotive force in a body, as arising from the motion of the body in the field, the alteration of the field itself, and the variation of electric potential from one part of the field to another.

(E) The relation between electric displacement, and the electromotive force which produces it.

(F) The relation between an electric current, and the electromotive force which produces it.

(G) The relation between the amount of free electricity at any point, and the electric displacements in the neighbourhood.

(H) The relation between the increase or diminution of free electricity and the electric currents in the neighbourhood.

There are twenty of these equations in all, involving twenty variable quantities.

(19) I then express in terms of these quantities the intrinsic energy of the Electromagnetic Field as depending partly on its magnetic and partly on its electric polarization at every point.

From this I determine the mechanical force acting, 1st, on a moveable conductor carrying an electric current; 2ndly, on a magnetic pole; 3rdly, on an electrified body.

The last result, namely, the mechanical force acting on an electrified body, gives rise to an independent method of electrical measurement founded on its electrostatic effects. The relation between the units employed in the two methods is shewn to depend on what I have called the "electric elasticity" of the medium, and to be a velocity, which has been experimentally determined by MM. Weber and Kohlrausch.

I then shew how to calculate the electrostatic capacity of a condenser, and the specific inductive capacity of a dielectric.

The case of a condenser composed of parallel layers of substances of different electric resistances and inductive capacities is next examined, and it is shewn that the phenomenon called electric absorption will generally occur, that is, the condenser, when suddenly discharged, will after a short time shew signs of a *residual* charge.

(20) The general equations are next applied to the case of a magnetic disturbance propagated through a non-conducting field, and it is shewn that the only disturbances which can be so propagated are those which are transverse to the direction of propagation, and that the velocity of propagation is the velocity $v$, found from experiments such as those of Weber, which expresses the number of electrostatic units of electricity which are contained in one electromagnetic unit.

This velocity is so nearly that of light, that it seems we have strong reason to conclude that light itself (including radiant heat, and other radiations if any) is an electromagnetic disturbance in the form of waves propagated through the electromagnetic field according to electromagnetic laws. If so, the agreement between the elasticity of the medium as calculated from the rapid alternations of luminous vibrations, and as found by the slow processes of electrical experiments, shews how perfect and regular the elastic properties of the medium must be when not encumbered with any matter denser than air. If the same character of the elasticity is retained in dense transparent bodies, it appears that the square of the index of refraction is equal to the product of the specific dielectric capacity and the specific magnetic capacity. Conducting media are shewn to absorb such radiations rapidly, and therefore to be generally opaque.

The conception of the propagation of transverse magnetic

disturbances to the exclusion of normal ones is distinctly set forth by Professor Faraday* in his "Thoughts on Ray Vibrations." The electromagnetic theory of light, as proposed by him, is the same in substance as that which I have begun to develope in this paper, except that in 1846 there were no data to calculate the velocity of propagation.

(21) The general equations are then applied to the calculation of the coefficients of mutual induction of two circular currents and the coefficient of self-induction, in a coil. The want of uniformity of the current in the different parts of the section of a wire at the commencement of the current is investigated, I believe for the first time, and the consequent correction of the coefficient of self-induction is found.

These results are applied to the calculation of the self-induction of the coil used in the experiments of the Committee of the British Association on Standards of Electric Resistance, and the value compared with that deduced from the experiments.

* *Philosophical Magazine*, May 1846, or *Experimental Researches*, III, p. 447.

PART II

# On Electromagnetic Induction

## Electromagnetic Momentum of Current

(22) We may begin by considering the state of the field in the neighbourhood of an electric current. We know that magnetic forces are excited in the field, their direction and magnitude depending according to known laws upon the form of the conductor carrying the current. When the strength of the current is increased, all the magnetic effects are increased in the same proportion. Now, if the magnetic state of the field depends on motions of the medium, a certain force must be exerted in order to increase or diminish these motions, and when the motions are excited they continue, so that the effect of the connexion between the current and the electromagnetic field surrounding it, is to endow the current with a kind of momentum, just as the connexion between the driving-point of a machine and a fly-wheel endows the driving-point with an additional momentum, which may be called the momentum of the fly-wheel reduced to the driving-point. The unbalanced force acting on the driving-point increases this momentum, and is measured by the rate of its increase.

In the case of electric currents, the resistance to sudden increase or diminution of strength produces effects exactly like those of momentum, but the amount of this momentum depends on the shape of the conductor and the relative position of its different parts.

## Mutual Action of two Currents

(23) If there are two electric currents in the field, the magnetic force at any point is that compounded of the forces due to each

current separately, and since the two currents are in connexion with every point of the field, they will be in connexion with each other, so that any increase or diminution of the one will produce a force acting with or contrary to the other.

### Dynamical Illustration of Reduced Momentum

(24) As a dynamical illustration, let us suppose a body $C$ so connected with two independent driving points $A$ and $B$ that its velocity is $p$ times that of $A$ together with $q$ times that of $B$. Let $u$ be the velocity of $A$, $v$ that of $B$, and $w$ that of $C$, and let $\delta x$, $\delta y$, $\delta z$ be their simultaneous displacements, then by the general equation of dynamics*,

$$C\frac{dw}{dt}\delta z = X\delta x + Y\delta y,$$

where $X$ and $Y$ are the forces acting at $A$ and $B$.

But

$$\frac{dw}{dt} = p\frac{du}{dt} + q\frac{dv}{dt},$$

and

$$\delta z = p\delta x + q\delta y.$$

Substituting, and remembering that $\delta x$ and $\delta y$ are independent,

$$\left.\begin{aligned} X &= \frac{d}{dt}(Cp^2u + Cpqv) \\ Y &= \frac{d}{dt}(Cpqu + Cq^2v) \end{aligned}\right\} \quad \ldots\ldots\ldots\ldots\ldots (1)$$

We may call $Cp^2u + Cpqv$ the momentum of $C$ referred to $A$, and $Cpqu + Cq^2v$ its momentum referred to $B$; then we may say that the effect of the force $X$ is to increase the momentum of $C$ referred to $A$, and that of $Y$ to increase its momentum referred to $B$.

If there are many bodies connected with $A$ and $B$ in a similar

* Lagrange, *Méc. Anal.* II. 2, § 5.

way but with different values of $p$ and $q$, we may treat the question in the same way by assuming

$$L = \Sigma\,(Cp^2), \quad M = \Sigma\,(Cpq), \quad \text{and } N = \Sigma\,(Cq^2),$$

where the summation is extended to all the bodies with their proper values of $C$, $p$, and $q$. Then the momentum of the system referred to $A$ is

$$Lu + Mv,$$

and referred to $B$,

$$Mu + Nv,$$

and we shall have

$$\left.\begin{aligned} X &= \frac{d}{dt}(Lu + Mv) \\ Y &= \frac{d}{dt}(Mu + Nv) \end{aligned}\right\} \ldots\ldots\ldots\ldots\ldots(2),$$

where $X$ and $Y$ are the external forces acting on $A$ and $B$.

(25) To make the illustration more complete we have only to suppose that the motion of $A$ is resisted by a force proportional to its velocity, which we may call $Ru$, and that of $B$ by a similar force, which we may call $Sv$, $R$ and $S$ being coefficients of resistance. Then if $\xi$ and $\eta$ are the forces on $A$ and $B$,

$$\left.\begin{aligned} \xi &= X + Ru = Ru + \frac{d}{dt}(Lu + Mv) \\ \eta &= Y + Sv = Sv + \frac{d}{dt}(Mu + Nv) \end{aligned}\right\} \ldots\ldots\ldots(3).$$

If the velocity of $A$ be increased at the rate $\frac{du}{dt}$, then in order to prevent $B$ from moving a force, $\eta = \frac{d}{dt}(Mu)$ must be applied to it.

This effect on $B$, due to an increase of the velocity of $A$, corresponds to the electromotive force on one circuit arising from an increase in the strength of a neighbouring circuit.

This dynamical illustration is to be considered merely as assisting the reader to understand what is meant in mechanics by Reduced Momentum. The facts of the induction of currents as depending on the variations of the quantity called Electromagnetic Momentum, or Electrotonic State, rest on the experiments of Faraday*, Felici†, &c.

## Coefficients of Induction for Two Circuits

(26) In the electromagnetic field the values of $L$, $M$, $N$ depend on the distribution of the magnetic effects due to the two circuits, and this distribution depends only on the form and relative position of the circuits. Hence $L$, $M$, $N$ are quantities depending on the form and relative position of the circuits, and are subject to variation with the motion of the conductors. It will be presently seen that $L$, $M$, $N$ are geometrical quantities of the nature of lines, that is, of one dimension in space; $L$ depends on the form of the first conductor, which we shall call $A$, $N$ on that of the second, which we shall call $B$, and $M$ on the relative position of $A$ and $B$.

(27) Let $\xi$ be the electromotive force acting on $A$, $x$ the strength of the current, and $R$ the resistance, then $Rx$ will be the resisting force. In steady currents the electromotive force just balances the resisting force, but in variable currents the resultant force $\xi - Rx$ is expended in increasing the "electromagnetic momentum," using the word momentum merely to express that which is generated by a force acting during a time, that is, a velocity existing in a body.

In the case of electric currents, the force in action is not ordinary mechanical force, at least we are not as yet able to measure it as common force, but we call it electromotive force, and the body moved is not merely the electricity in the conductor, but something outside the conductor, and capable of being affected by other conductors in the neighbourhood carrying currents. In this it resembles rather the reduced momentum of a driving-point of a

* *Experimental Researches*, Series I., IX.

† *Annales de Chimie*, sér. 3, XXXIV. (1852), p. 64.

machine as influenced by its mechanical connexions, than that of a simple moving body like a cannon ball, or water in a tube.

### Electromagnetic Relations of two Conducting Circuits

(28) In the case of two conducting circuits, $A$ and $B$, we shall assume that the electromagnetic momentum belonging to $A$ is

$$Lx + My,$$

and that belonging to $B$,

$$Mx + Ny,$$

where $L$, $M$, $N$ correspond to the same quantities in the dynamical illustration, except that they are supposed to be capable of variation when the conductors $A$ or $B$ are moved.

Then the equation of the current $x$ in $A$ will be

$$\xi = Rx + \frac{d}{dt}(Lx + My) \quad \ldots\ldots\ldots\ldots\ldots (4),$$

and that of $y$ in $B$

$$\eta = Sy + \frac{d}{dt}(Mx + Ny) \quad \ldots\ldots\ldots\ldots\ldots (5),$$

where $\xi$ and $\eta$ are the electromotive forces, $x$ and $y$ the currents, and $R$ and $S$ the resistances in $A$ and $B$ respectively.

### Induction of one Current by another

(29) Case 1st. Let there be no electromotive force on $B$, except that which arises from the action of $A$, and let the current of $A$ increase from 0 to the value $x$, then

$$Sy + \frac{d}{dt}(Mx + Ny) = 0,$$

whence

$$Y = \int_0^t y\,dt = -\frac{M}{S}x, \quad \ldots\ldots\ldots\ldots\ldots (6)$$

that is, a quantity of electricity $Y$, being the total induced current, will flow through $B$ when $x$ rises from 0 to $x$. This is induction by variation of the current in the primary conductor. When $M$ is positive, the induced current due to increase of the primary current is negative.

### Induction by Motion of Conductor

(30) Case 2nd. Let $x$ remain constant, and let $M$ change from $M$ to $M'$, then

$$Y = -\frac{M' - M}{S} x; \quad \ldots\ldots\ldots\ldots\ldots (7)$$

so that if $M$ is increased, which it will be by the primary and secondary circuits approaching each other, there will be a negative induced current, the total quantity of electricity passed through $B$ being $Y$.

This is induction by the relative motion of the primary and secondary conductors.

### Equation of Work and Energy

(31) To form the equation between work done and energy produced, multiply (1) by $x$ and (2) by $y$, and add

$$\xi x + \eta y = Rx^2 + Sy^2 + x\frac{d}{dt}(Lx + My) + y\frac{d}{dt}(Mx + Ny) \quad \ldots\ldots (8)$$

Here $\xi x$ is the work done in unit of time by the electromotive force $\xi$ acting on the current $x$ and maintaining it, and $\eta y$ is the work done by the electromotive force $\eta$. Hence the left-hand side of the equation represents the work done by the electromotive forces in unit of time.

### Heat produced by the Current

(32) On the other side of the equation we have, first,

$$Rx^2 + Sy^2 = H \ldots\ldots\ldots\ldots\ldots\ldots (9),$$

which represents the work done in overcoming the resistance of the circuits in unit of time. This is converted into heat. The remaining terms represent work not converted into heat. They may be written

$$\tfrac{1}{2}\frac{d}{dt}(Lx^2 + 2Mxy + Ny^2) + \tfrac{1}{2}\frac{dL}{dt}x^2 + \frac{dM}{dt}xy + \tfrac{1}{2}\frac{dN}{dt}y^2.$$

### Intrinsic Energy of the Currents

(33) If $L$, $M$, $N$ are constant, the whole work of the electromotive forces which is not spent against resistance will be devoted to the development of the currents. The whole intrinsic energy of the currents is therefore

$$\tfrac{1}{2}Lx^2 + Mxy + \tfrac{1}{2}Ny^2 = E \ldots\ldots\ldots\ldots (10).$$

This energy exists in a form imperceptible to our senses, probably as actual motion, the seat of this motion being not merely the conducting circuits, but the space surrounding them.

### Mechanical Action between Conductors

(34) The remaining terms,

$$\tfrac{1}{2}\frac{dL}{dt}x^2 + \frac{dM}{dt}xy + \tfrac{1}{2}\frac{dN}{dt}y^2 = W \quad \ldots\ldots\ldots (11),$$

represent the work done in unit of time arising from the variations of $L$, $M$, and $N$, or, what is the same thing, alterations in the form and position of the conducting circuits $A$ and $B$.

Now if work is done when a body is moved, it must arise from ordinary mechanical force acting on the body while it is moved. Hence this part of the expression shews that there is a mechanical force urging every part of the conductors themselves in that direction in which $L$, $M$, and $N$ will be most increased.

The existence of the electromagnetic force between conductors carrying currents is therefore a direct consequence of the joint and independent action of each current on the electromagnetic field. If $A$ and $B$ are allowed to approach a distance $ds$, so as to increase

$M$ from $M$ to $M'$ while the currents are $x$ and $y$, then the work done will be

$$(M' - M)\,xy,$$

and the force in the direction of $ds$ will be

$$\frac{dM}{ds}xy \quad \ldots\ldots\ldots\ldots\ldots\ldots\ldots (12),$$

and this will be an attraction if $x$ and $y$ are of the same sign, and if $M$ is increased as $A$ and $B$ approach.

It appears, therefore, that if we admit that the unresisted part of electromotive force goes on as long as it acts, generating a self-persistent state of the current, which we may call (from mechanical analogy) its electromagnetic momentum, and that this momentum depends on circumstances external to the conductor, then both induction of currents and electromagnetic attractions may be proved by mechanical reasoning.

What I have called electromagnetic momentum is the same quantity which is called by Faraday* the electrotonic state of the circuit, every change of which involves the action of an electromotive force, just as change of momentum involves the action of mechanical force.

If, therefore, the phenomena described by Faraday in the Ninth Series of his *Experimental Researches* were the only known facts about electric currents, the laws of Ampère relating to the attraction of conductors carrying currents, as well as those of Faraday about the mutual induction of currents, might be deduced by mechanical reasoning.

In order to bring these results within the range of experimental verification, I shall next investigate the case of a single current, of two currents, and of the six currents in the electric balance, so as to enable the experimenter to determine the values of $L$, $M$, $N$.

### Case of a single Circuit

(35) The equation of the current $x$ in a circuit whose resistance is $R$, and whose coefficient of self-induction is $L$, acted on by an

* *Experimental Researches*, Series I. 60, &c.

external electromotive force $\xi$, is

$$\xi - Rx = \frac{d}{dt} Lx \quad \ldots\ldots\ldots\ldots\ldots\ldots (13).$$

When $\xi$ is constant, the solution is of the form

$$x = b + (a - b)\, e^{-\frac{L}{R}t},$$

where $a$ is the value of the current at the commencement, and $b$ is its final value.

The total quantity of electricity which passes in time $t$, where $t$ is great, is

$$\int_0^t x dt = bt + (a - b)\frac{L}{R} \quad \ldots\ldots\ldots\ldots\ldots\ldots (14).$$

The value of the integral of $x^2$ with respect to the time is

$$\int_0^t x^2 dt = b^2 t + (a - b)\frac{L}{R}\left(\frac{3b + a}{2}\right) \quad \ldots\ldots\ldots (15).$$

The actual current changes gradually from the initial value $a$ to the final value $b$, but the values of the integrals of $x$ and $x^2$ are the same as if a steady current of intensity $\frac{1}{2}(a + b)$ were to flow for a time $2\frac{L}{R}$, and were then succeeded by the steady current $b$. The time $2\frac{L}{R}$ is generally so minute a fraction of a second, that the effects on the galvanometer and dynamometer may be calculated as if the impulse were instantaneous.

If the circuit consists of a battery and a coil, then, when the circuit is first completed, the effects are the same as if the current had only half its final strength during the time $2\frac{L}{R}$. This diminution of the current, due to induction, is sometimes called the counter-current.

(36) If an additional resistance $r$ is suddenly thrown into the circuit, as by breaking contact, so as to force the current to pass

through a thin wire of resistance $r$, then the original current is $a = \frac{\xi}{R}$, and the final current is $b = \frac{\xi}{R+r}$.

The current of induction is then $\frac{1}{2}\xi \frac{2R+r}{R(R+r)}$, and continues for a time $2\frac{L}{R+r}$. This current is greater than that which the battery can maintain in the two wires $R$ and $r$, and may be sufficient to ignite the thin wire $r$.

When contact is broken by separating the wires in air, this additional resistance is given by the interposed air, and since the electromotive force across the new resistance is very great, a spark will be forced across.

If the electromotive force is of the form $E \sin pt$, as in the case of a coil revolving in the magnetic field, then

$$x = \frac{E}{\rho} \sin (pt - \alpha),$$

where $\rho^2 = R^2 + L^2p^2$, and $\tan \alpha = \frac{Lp}{R}$.

## Case of two Circuits

(37) Let $R$ be the primary circuit and $S$ the secondary circuit then we have a case similar to that of the induction coil.

The equations of currents are those marked $A$ and $B$, and we may here assume $L$, $M$, $N$ as constant because there is no motion of the conductors. The equations then become

$$\left.\begin{aligned} Rx + L\frac{dx}{dt} + M\frac{dy}{dt} &= \xi \\ Sy + M\frac{dx}{dt} + N\frac{dy}{dt} &= 0 \end{aligned}\right\} \quad \ldots\ldots\ldots\ldots(13^*).$$

To find the total quantity of electricity which passes, we have only to integrate these equations with respect to $t$; then if $x_0$, $y_0$

be the strengths of the currents at time 0, and $x_1$, $y_1$, at time $t$, and if $X$, $Y$ be the quantities of electricity passed through each circuit during time $t$,

$$\left.\begin{aligned} X &= \frac{1}{R}\{\xi t + L(x_0 - x_1) + M(y_0 - y_1)\} \\ Y &= \frac{1}{S}\{M(x_0 - x_1) + N(y_0 - y_1)\} \end{aligned}\right\} \quad \ldots(14^*).$$

When the circuit $R$ is completed, then the total currents up to time $t$, when $t$ is great, are found by making

$$x_0 = 0, \quad x_1 = \frac{\xi}{R}, \quad y_0 = 0, \quad y_1 = 0;$$

then

$$X = x_1\left(t - \frac{L}{R}\right), \quad Y = -\frac{M}{S}x_1 \quad \ldots\ldots\ldots(15^*)$$

The value of the total counter-current in $R$ is therefore independent of the secondary circuit, and the induction current in the secondary circuit depends only on $M$, the coefficient of induction between the coils, $S$ the resistance of the secondary coil, and $x_1$ the final strength of the current in $R$.

When the electromotive force $\xi$ ceases to act, there is an extra current in the primary circuit, and a positive induced current in the secondary circuit, whose values are equal and opposite to those produced on making contact.

(38) All questions relating to the total quantity of transient currents, as measured by the impulse given to the magnet of the galvanometer, may be solved in this way without the necessity of a complete solution of the equations. The heating effect of the current, and the impulse it gives to the suspended coil of Weber's dynamometer, depend on the square of the current at every instant during the short time it lasts. Hence we must obtain the solution of the equations, and from the solution we may find the effects both on the galvanometer and dynamometer; and we may then make use of the method of Weber for estimating the intensity and

duration of a current uniform while it lasts which would produce the same effects.

(39) Let $n_1$, $n_2$ be the roots of the equation

$$(LN - M^2)\,n^2 + (RN + LS)\,n + RS = 0 \quad \ldots\ldots(16)$$

and let the primary coil be acted on by a constant electromotive force $Rc$, so that $c$ is the constant current it could maintain; then the complete solution of the equations for making contact is

$$x = \frac{c}{S}\frac{n_1 n_2}{n_1 - n_2}\left\{\left(\frac{S}{n_1} + N\right)e^{n_1 t} - \left(\frac{S}{n_2} + N\right)e^{n_2 t} + S\frac{n_1 - n^2}{n_1 n_2}\right\} \quad \ldots\ldots(17),$$

$$y = \frac{cM}{S}\frac{n_1 n_2}{n_1 - n_2}\{e^{n_1 t} - e^{n_2 t}\} \quad \ldots\ldots(18),$$

From these we obtain for calculating the impulse on the dynamometer,

$$\int x^2 dt = c^2\left\{t - \tfrac{3}{2}\frac{L}{R} - \tfrac{1}{2}\frac{M^2}{RN + LS}\right\} \quad \ldots\ldots(19),$$

$$\int y^2 dt = c^2\,\tfrac{1}{2}\frac{M^2 R}{S(RN + LS)} \quad \ldots\ldots(20).$$

The effects of the current in the secondary coil on the galvanometer and dynamometer are the same as those of a uniform current

$$-\tfrac{1}{2}c\frac{MR}{RN + LS}$$

for a time

$$2\left(\frac{L}{R} + \frac{N}{S}\right).$$

(40) The equation between work and energy may be easily verified. The work done by the electromotive force is

$$\xi\int x\,dt = c^2\,(Rt - L).$$

Work done in overcoming resistance and producing heat,

$$R\int x^2 dt + S\int y^2 dt = c^2\,(Rt - \tfrac{3}{2}L).$$

Energy remaining in the system,

$$= \tfrac{1}{2}c^2 L.$$

(41) If the circuit $R$ is suddenly and completely interrupted while carrying a current $c$, then the equation of the current in the secondary coil would be

$$y = c\frac{M}{N}e^{-\frac{S}{N}t}.$$

This current begins with a value $c\frac{M}{N}$, and gradually disappears.

The total quantity of electricity is $c\frac{M}{S}$, and the value of $\int y^2 dt$ is $c^2\frac{M^2}{2SN}$.

The effects on the galvanometer and dynamometer are equal to those of a uniform current $\tfrac{1}{2}c\frac{M}{N}$ for a time $2\frac{N}{S}$.

The heating effect is therefore greater than that of the current on making contact.

(42) If an electromotive force of the form $\xi = E\cos pt$ acts on the circuit $R$, then if the circuit $S$ is removed, the value of $x$ will be

$$x = \frac{E}{A}\sin(pt - \alpha),$$

where

$$A^2 = R^2 + L^2p^2,$$

and

$$\tan\alpha = \frac{Lp}{R}.$$

The effect of the presence of the circuit $S$ in the neighbourhood is to alter the value of $A$ and $\alpha$, to that which they would be if $R$ became

$$R + p^2\frac{MS}{S^2 + p^2N^2},$$

and $L$ became
$$L - p^2 \frac{MN}{S^2 + p^2 N^2}.$$

Hence the effect of the presence of the circuit $S$ is to increase the apparent resistance and diminish the apparent self-induction of the circuit $R$.

[*Paragraphs* (43)–(46), *which have been omitted, deal with the determination of coefficients of self and mutual inductance, by electrical bridge experiments.*]

## Exploration of the Electromagnetic Field

(47) Let us now suppose the primary circuit $A$ to be of invariable form, and let us explore the electromagnetic field by means of the secondary circuit $B$, which we shall suppose to be variable in form and position.

We may begin by supposing $B$ to consist of a short straight conductor with its extremities sliding on two parallel conducting rails, which are put in connexion at some distance from the sliding-piece.

Then, if sliding the moveable conductor in a given direction increases the value of $M$, a negative electromotive force will act in the circuit $B$, tending to produce a negative current in $B$ during the motion of the sliding-piece.

If a current be kept up in the circuit $B$, then the sliding-piece will itself tend to move in that direction, which causes $M$ to increase. At every point of the field there will always be a certain direction such that a conductor moved in that direction does not experience any electromotive force in whatever direction its extremities are turned. A conductor carrying a current will experience no mechanical force urging it in that direction or the opposite.

This direction is called the direction of the line of magnetic force through that point.

Motion of a conductor across such a line produces electromotive force in a direction perpendicular to the line and to the direction of motion, and a conductor carrying a current is urged in a direction perpendicular to the line and to the direction of the current.

(48) We may next suppose $B$ to consist of a very small plane circuit capable of being placed in any position and of having its plane turned in any direction. The value of $M$ will be greatest when the plane of the circuit is perpendicular to the line of magnetic force. Hence if a current is maintained in $B$ it will tend to set itself in this position, and will of itself indicate, like a magnet, the direction of the magnetic force.

## On Lines of Magnetic Force

(49) Let any surface be drawn, cutting the lines of magnetic force, and on this surface let any system of lines be drawn at small intervals, so as to lie side by side without cutting each other. Next, let any line be drawn on the surface cutting all these lines, and let a second line be drawn near it, its distance from the first being such that the value of $M$ for each of the small spaces enclosed between these two lines and the lines of the first system is equal to unity.

In this way let more lines be drawn so as to form a second system, so that the value of $M$ for every reticulation formed by the intersection of the two systems of lines is unity.

Finally, from every point of intersection of these reticulations let a line be drawn through the field, always coinciding in direction with the direction of magnetic force.

(50) In this way the whole field will be filled with lines of magnetic force at regular intervals, and the properties of the electromagnetic field will be completely expressed by them.

For, 1st, If any closed curve be drawn in the field, the value of $M$ for that curve will be expressed by the *number* of lines of force which *pass through* that closed curve.

2ndly. If this curve be a conducting circuit and be moved through the field, an electromotive force will act in it, represented by the rate of decrease of the number of lines passing through the curve.

3rdly. If a current be maintained in the circuit, the conductor will be acted on by forces tending to move it so as to increase the number of lines passing through it, and the amount of work done by these forces is equal to the current in the circuit multiplied by the number of additional lines.

4thly. If a small plane circuit be placed in the field, and be free to turn, it will place its plane perpendicular to the lines of force. A small magnet will place itself with its axis in the direction of the lines of force.

5thly. If a long uniformly magnetized bar is placed in the field, each pole will be acted on by a force in the direction of the lines of force. The number of lines of force passing through unit of area is equal to the force acting on a unit pole multiplied by a coefficient depending on the magnetic nature of the medium, and called the coefficient of magnetic induction.

In fluids and isotropic solids the value of this coefficient $\mu$ is the same in whatever direction the lines of force pass through the substance, but in crystallized, strained, and organized solids the value of $\mu$ may depend on the direction of the lines of force with respect to the axes of crystallization, strain, or growth.

In all bodies $\mu$ is affected by temperature, and in iron it appears to diminish as the intensity of the magnetization increases.

## On Magnetic Equipotential Surfaces

(51) If we explore the field with a uniformly magnetized bar, so long that one of its poles is in a very weak part of the magnetic field, then the magnetic forces will perform work on the other pole as it moves about the field.

If we start from a given point, and move this pole from it to any other point, the work performed will be independent of the path of the pole between the two points; provided that no electric current passes between the different paths pursued by the pole.

Hence, when there are no electric currents but only magnets in the field, we may draw a series of surfaces such that the work done in passing from one to another shall be constant whatever be the path pursued between them. Such surfaces are called Equipotential Surfaces, and in ordinary cases are perpendicular to the Lines of magnetic force.

If these surfaces are so drawn that, when a unit pole passes from any one to the next in order, unity of work is done, then the work done in any motion of a magnetic pole will be measured by the strength of the pole multiplied by the number of surfaces which it has passed through in the positive direction.

(52) If there are circuits carrying electric currents in the field, then there will still be equipotential surfaces in the parts of the field external to the conductors carrying the currents, but the work done on a unit pole in passing from one to another will depend on the number of times which the path of the pole circulates round any of these currents. Hence the potential in each surface will have a series of values in arithmetical progression, differing by the work done in passing completely round one of the currents in the field.

The equipotential surfaces will not be continuous closed surfaces, but some of them will be limited sheets, terminating in the electric circuit as their common edge or boundary. The number of these will be equal to the amount of work done on a unit pole in going round the current, and this by the ordinary measurement $= 4\pi\gamma$, where $\gamma$ is the value of the current.

These surfaces, therefore, are connected with the electric current as soap-bubbles are connected with a ring in M. Plateau's experiments. Every current $\gamma$ has $4\pi\gamma$ surfaces attached to it. These surfaces have the current for their common edge, and meet it at equal angles. The form of the surfaces in other parts depends on the presence of other currents and magnets, as well as on the shape of the circuit to which they belong.

PART III

# General Equations of the Electromagnetic Field

(53) Let us assume three rectangular directions in space as the axes of $x$, $y$, and $z$, and let all quantities having direction be expressed by their components in these three directions.

## Electrical Currents (p, q, r)

(54) An electrical current consists in the transmission of electricity from one part of a body to another. Let the quantity of electricity transmitted in unit of time across unit of area perpendicular to the axis of $x$ be called $p$, then $p$ is the component of the current at that place in the direction of $x$.

We shall use the letters $p$, $q$, $r$ to denote the components of the current per unit of area in the directions of $x$, $y$, $z$.

## Electrical Displacements (f, g, h)

(55) Electrical displacement consists in the opposite electrification of the sides of a molecule or particle of a body which may or may not be accompanied with transmission through the body. Let the quantity of electricity which would appear on the faces $dy.dz$ of an element $dx$, $dy$, $dz$ cut from the body be $f.dy.dz$, then $f$ is the component of electric displacement parallel to $x$. We shall use $f$, $g$, $h$ to denote the electric displacements parallel to $x$, $y$, $z$ respectively.

The variations of the electrical displacement must be added to the currents $p$, $q$, $r$ to get the total motion of electricity, which we may call $p'$, $q'$, $r'$, so that

$$\left.\begin{aligned} p' &= p + \frac{df}{dt} \\ q' &= q + \frac{dg}{dt} \\ r' &= r + \frac{dh}{dt} \end{aligned}\right\} \quad \ldots\ldots\ldots\ldots\ldots\ldots (A).$$

## Electromotive Force (P, Q, R)

(56) Let $P$, $Q$, $R$ represent the components of the electromotive force at any point. Then $P$ represents the difference of potential per unit of length in a conductor placed in the direction of $x$ at the given point. We may suppose an indefinitely short wire placed parallel to $x$ at a given point and touched, during the action of the force $P$, by two small conductors, which are then insulated and removed from the influence of the electromotive force. The value of $P$ might then be ascertained by measuring the charge of the conductors.

Thus if $l$ be the length of the wire, the difference of potential at its ends will be $Pl$, and if $C$ be the capacity of each of the small conductors the charge on each will be $\frac{1}{2}CPl$. Since the capacities of moderately large conductors, measured on the electromagnetic system, are exceedingly small, ordinary electromotive forces arising from electromagnetic actions could hardly be measured in this way. In practice such measurements are always made with long conductors, forming closed or nearly closed circuits.

## Electromagnetic Momentum (F, G, H)

(57) Let $F$, $G$, $H$ represent the components of electromagnetic momentum at any point of the field, due to any system of magnets or currents.

Then $F$ is the total impulse of the electromotive force in the direction of $x$ that would be generated by the removal of these magnets or currents from the field, that is, if $P$ be the electromotive force at any instant during the removal of the system

$$F = \int P dt.$$

Hence the part of the electromotive force which depends on the motion of magnets or currents in the field, or their alteration of intensity, is

$$P = -\frac{dF}{dt}, \quad Q = -\frac{dG}{dt}, \quad R = -\frac{dH}{dt} \quad \ldots\ldots\ldots (29).$$

### Electromagnetic Momentum of a Circuit

(58) Let $s$ be the length of the circuit, then if we integrate

$$\int \left( F\frac{dx}{ds} + G\frac{dy}{ds} + H\frac{dz}{ds} \right) ds \quad \ldots\ldots\ldots\ldots (30)$$

round the circuit, we shall get the total electromagnetic momentum of the circuit, or the number of lines of magnetic force which pass through it, the variations of which measure the total electromotive force in the circuit. This electromagnetic momentum is the same thing to which Professor Faraday has applied the name of the Electrotonic State.

If the circuit be the boundary of the elementary area $dy\,dz$, then its electromagnetic momentum is

$$\left( \frac{dH}{dy} - \frac{dG}{dz} \right) dy\,dz,$$

and this is the number of lines of magnetic force which pass through the area $dy\,dz$.

### Magnetic Force ($\alpha$, $\beta$, $\gamma$)

(59) Let $\alpha$, $\beta$, $\gamma$ represent the force acting on a unit magnetic pole placed at the given point resolved in the directions of $x$, $y$, and $z$.

## Coefficient of Magnetic Induction ($\mu$)

(60) Let $\mu$ be the ratio of the magnetic induction in a given medium to that in air under an equal magnetizing force, then the number of lines of force in unit of area perpendicular to $x$ will be $\mu\alpha$ ($\mu$ is a quantity depending on the nature of the medium, its temperature, the amount of magnetization already produced, and in crystalline bodies varying with the direction).

(61) Expressing the electric momentum of small circuits perpendicular to the three axes in this notation, we obtain the following

## Equations of Magnetic Force

$$\left.\begin{aligned} \mu\alpha &= \frac{dH}{dy} - \frac{dG}{dz} \\ \mu\beta &= \frac{dF}{dz} - \frac{dH}{dx} \\ \mu\gamma &= \frac{dG}{dx} - \frac{dF}{dy} \end{aligned}\right\} \quad \ldots\ldots\ldots\ldots\ldots (B).$$

## Equations of Currents

(62) It is known from experiment that the motion of a magnetic pole in the electromagnetic field in a closed circuit cannot generate work unless the circuit which the pole describes passes round an electric current. Hence, except in the space occupied by the electric currents,

$$\alpha dx + \beta dy + \gamma dz = d\phi \quad \ldots\ldots\ldots\ldots\ldots (31)$$

a complete differential of $\phi$, the magnetic potential.

The quantity $\phi$ may be susceptible of an indefinite number of distinct values, according to the number of times that the exploring point passes round electric currents in its course, the difference between successive values of $\phi$ corresponding to a passage completely round a current of strength $c$ being $4\pi c$.

Hence if there is no electric current,

$$\frac{d\gamma}{dy} - \frac{d\beta}{dz} = 0;$$

but if there is a current $p'$,

Similarly,

$$\left.\begin{aligned} \frac{d\gamma}{dy} - \frac{d\beta}{dz} &= 4\pi p' \\ \frac{d\alpha}{dz} - \frac{d\gamma}{dx} &= 4\pi q' \\ \frac{d\beta}{dx} - \frac{d\alpha}{dy} &= 4\pi r' \end{aligned}\right\} \quad \ldots\ldots\ldots\ldots\ldots (C).$$

We may call these the Equations of Currents.

### Electromotive Force in a Circuit

(63) Let $\xi$ be the electromotive force acting round the circuit $A$, then

$$\xi = \int \left( P\frac{dx}{ds} + Q\frac{dy}{ds} + R\frac{dz}{ds} \right) ds \quad \ldots\ldots\ldots (32),$$

where $ds$ is the element of length, and the integration is performed round the circuit.

Let the forces in the field be those due to the circuits $A$ and $B$, then the electromagnetic momentum of $A$ is

$$\int \left( F\frac{dx}{ds} + G\frac{dy}{ds} + H\frac{dz}{ds} \right) ds = Lu + Mv \quad \ldots\ldots (33),$$

where $u$ and $v$ are the currents in $A$ and $B$, and

$$\xi = -\frac{d}{dt}(Lu + Mv) \quad \ldots\ldots\ldots\ldots\ldots (34).$$

Hence, if there is no motion of the circuit $A$,

$$\left.\begin{aligned} P &= -\frac{dF}{dt} - \frac{d\psi}{dx} \\ Q &= -\frac{dG}{dt} - \frac{d\psi}{dy} \\ R &= -\frac{dH}{dt} - \frac{d\psi}{dz} \end{aligned}\right\} \quad \ldots\ldots\ldots\ldots\ldots (35),$$

where $\psi$ is a function of $x$, $y$, $z$, and $t$, which is indeterminate as far as regards the solution of the above equations, because the terms depending on it will disappear on integrating round the circuit. The quantity $\psi$ can always, however, be determined in any particular case when we know the actual conditions of the question. The physical interpretation of $\psi$ is, that it represents the *electric potential* at each point of space.

## Electromotive Force on a Moving Conductor

(64) Let a short straight conductor of length $a$, parallel to the axis of $x$, move with a velocity whose components are $\frac{dx}{dt}, \frac{dy}{dt}, \frac{dz}{dt}$, and let its extremities slide along two parallel conductors with a velocity $\frac{ds}{dt}$. Let us find the alteration of the electromagnetic momentum of the circuit of which this arrangement forms a part.

In unit of time the moving conductor has travelled distances $\frac{dx}{dt}, \frac{dy}{dt}, \frac{dz}{dt}$ along the directions of the three axes, and at the same time the lengths of the parallel conductors included in the circuit have each been increased by $\frac{ds}{dt}$.

Hence the quantity

$$\int\left(F\frac{dx}{ds}+G\frac{dy}{ds}+H\frac{dz}{ds}\right)ds$$

will be increased by the following increments,

$$a\left(\frac{dF}{dx}\frac{dx}{dt}+\frac{dF}{dy}\frac{dy}{dt}+\frac{dF}{dz}\frac{dz}{dt}\right),\ \text{due to motion of conductor,}$$

$$-a\frac{ds}{dt}\left(\frac{dF}{dx}\frac{dx}{ds}+\frac{dG}{dx}\frac{dy}{ds}+\frac{dH}{dx}\frac{dz}{ds}\right),\ \text{due to lengthening of circuit.}$$

The total increment will therefore be

$$a\left(\frac{dF}{dy}-\frac{dG}{dx}\right)\frac{dy}{dt}-a\left(\frac{dH}{dx}-\frac{dF}{dz}\right)\frac{dz}{dt};$$

or, by the equations of Magnetic Force (8),

$$-a\left(\mu\gamma\frac{dy}{dt}-\mu\beta\frac{dz}{dt}\right).$$

If $P$ is the electromotive force in the moving conductor parallel to $x$ referred to unit of length, then the actual electromotive force is $Pa$; and since this is measured by the decrement of the electromagnetic momentum of the circuit, the electromotive force due to motion will be

$$P=\mu\gamma\frac{dy}{dt}-\mu\beta\frac{dz}{dt} \quad \ldots\ldots\ldots\ldots\ldots(36).$$

(65) The complete equations of electromotive force on a moving conductor may now be written as follows:—

**Equations of Electromotive Force**

$$\left.\begin{aligned} P&=\mu\left(\gamma\frac{dy}{dt}-\beta\frac{dz}{dt}\right)-\frac{dF}{dt}-\frac{d\psi}{dx}\\ Q&=\mu\left(\alpha\frac{dz}{dt}-\gamma\frac{dx}{dt}\right)-\frac{dG}{dt}-\frac{d\psi}{dy}\\ R&=\mu\left(\beta\frac{dx}{dt}-\alpha\frac{dy}{dt}\right)-\frac{dH}{dt}-\frac{d\psi}{dz}\end{aligned}\right\} \ldots\ldots(D).$$

The first term on the right-hand side of each equation represents the electromotive force arising from the motion of the conductor itself. This electromotive force is perpendicular to the direction of motion and to the lines of magnetic force; and if a parellelogram be drawn whose sides represent in direction and magnitude the velocity of the conductor and the magnetic induction at that point of the field, then the area of the parallelogram will represent the electromotive force due to the motion of the conductor, and the direction of the force is perpendicular to the plane of the parellogram.

The second term in each equation indicates the effect of changes in the position or strength of magnets or currents in the field.

The third term shews the effect of the electric potential $\psi$. It has no effect in causing a circulating current in a closed circuit. It indicates the existence of a force urging the electricity to or from certain definite points in the field.

### Electric Elasticity

(66) When an electromotive force acts on a dielectric, it puts every part of the dielectric into a polarized condition, in which its opposite sides are oppositely electrified. The amount of this electrification depends on the electromotive force and on the nature of the substance, and, in solids having a structure defined by axes, on the direction of the electromotive force with respect to these axes. In isotropic substances, if $k$ is the ratio of the electromotive force to the electric displacement, we may write the

### Equations of Electric Elasticity

$$\left.\begin{aligned} P &= kf \\ Q &= kg \\ R &= kh \end{aligned}\right\} \quad \ldots\ldots\ldots\ldots\ldots\ldots (E).$$

### Electric Resistance

(67) When an electromotive force acts on a conductor it produces a current of electricity through it. This effect is additional to the electric displacement already considered. In solids of complex structure, the relation between the electromotive force and the current depends on their direction through the solid. In isotropic substances, which alone we shall here consider, if $\rho$ is the specific resistance referred to unit of volume, we may write the

### Equations of Electric Resistance

$$\left.\begin{aligned} P &= -\rho p \\ Q &= -\rho q \\ R &= -\rho r \end{aligned}\right\} \quad \ldots\ldots\ldots\ldots\ldots\ldots (F).$$

### Electric Quantity

(68) Let $e$ represent the quantity of free positive electricity contained in unit of volume at any part of the field, then, since this arises from the electrification of the different parts of the field not neutralizing each other, we may write the

### Equation of Free Electricity

$$e + \frac{df}{dx} + \frac{dg}{dy} + \frac{dh}{dz} = 0 \quad \ldots\ldots\ldots\ldots\ldots (G).$$

(69) If the medium conducts electricity, then we shall have another condition, which may be called, as in hydrodynamics, the

### Equation of Continuity

$$\frac{de}{dt} + \frac{dp}{dx} + \frac{dq}{dy} + \frac{dr}{dz} = 0 \quad \ldots\ldots\ldots\ldots\ldots (H).$$

(70) In these equations of the electromagnetic field we have assumed twenty variable quantities, namely,

| | | | |
|---|---|---|---|
| For Electromagnetic Momentum ......... | F | G | H |
| ,, Magnetic Intensity .................. | $\alpha$ | $\beta$ | $\gamma$ |
| ,, Electromotive Force ................ | $P$ | $Q$ | $R$ |
| ,, Current due to true Conduction...... | $p$ | $q$ | $r$ |
| ,, Electric Displacement............... | $f$ | $g$ | $h$ |
| ,, Total Current (including variation of displacement).................... | $p'$ | $q'$ | $r'$ |
| ,, Quantity of Free Electricity.......... | $e$ | | |
| ,, Electric Potential.................. | $\Psi$ | | |

Between these twenty quantities we have found twenty equations, viz.

Three equations of Magnetic Force . . . . . . . . . . . . . . (B)
" Electric Currents . . . . . . . . . . . . . . (C)
" Electromotive Force . . . . . . . . . . . (D)
" Electric Elasticity . . . . . . . . . . . . . (E)
" Electric Resistance . . . . . . . . . . . . (F)
" Total Currents . . . . . . . . . . . . . . . . (A)
One equation of Free Electricity . . . . . . . . . . . . . . . . . (G)
" Continuity . . . . . . . . . . . . . . . . . . . . . (H)

These equations are therefore sufficient to determine all the quantities which occur in them, provided we know the conditions of the problem. In many questions, however, only a few of the equations are required.

## Intrinsic Energy of the Electromagnetic Field

(71) We have seen (33) that the intrinsic energy of any system of currents is found by multiplying half the current in each circuit into its electromagnetic momentum. This is equivalent to finding the integral

$$E = \tfrac{1}{2}\Sigma(Fp' + Gq' + Hr')dV \qquad \ldots\ldots\ldots (37)$$

over all the space occupied by currents, where $p$, $q$, $r$ are the components of currents, and $F$, $G$, $H$ the components of electromagnetic momentum.

Substituting the values of $p'$, $q'$ $r'$ from the equations of Currents (C), this becomes

$$\frac{1}{8\pi}\Sigma\left\{F\left(\frac{d\gamma}{dy}-\frac{d\beta}{dz}\right)+G\left(\frac{d\alpha}{dz}-\frac{d\gamma}{dx}\right)+H\left(\frac{d\beta}{dx}-\frac{d\alpha}{dy}\right)\right\}dV.$$

Integrating by parts, and remembering that $\alpha$, $\beta$, $\gamma$ vanish at an infinite distance, the expression becomes

$$\frac{1}{8\pi}\Sigma\left\{\alpha\left(\frac{dH}{dy}-\frac{dG}{dz}\right)+\beta\left(\frac{dF}{dz}-\frac{dH}{dx}\right)+\gamma\left(\frac{dG}{dx}-\frac{dF}{dy}\right)\right\}dV,$$

where the integration is to be extended over all space. Referring to the equations of Magnetic Force (B), p. 258, this becomes

$$E=\frac{1}{8\pi}\Sigma\{\alpha.\mu\alpha+\beta.\mu\beta+\gamma.\mu\gamma\}\,dV \ldots\ldots\ldots (38),$$

where $\alpha$, $\beta$, $\gamma$ are the components of magnetic intensity or the force on a unit magnetic pole, and $\mu\alpha$, $\mu\beta$, $\mu\gamma$ are the components of the quantity of magnetic induction, or the number of lines of force in unit of area.

In isotropic media the value of $\mu$ is the same in all directions, and we may express the result more simply by saying that the intrinsic energy of any part of the magnetic field arising from its magnetization is

$$\frac{\mu}{8\pi}I^2$$

per unit of volume, where $I$ is the magnetic intensity.

(72) Energy may be stored up in the field in a different way, namely, by the action of electromotive force in producing electric displacement. The work done by a variable electromotive force, $P$, in producing a variable displacement, $f$, is got by integrating

$$\int P df$$

from $P = 0$ to the given value of $P$.

Since $P = kf$, equation (E), this quantity becomes

$$\int kf df = \tfrac{1}{2}kf^2 = \tfrac{1}{2}Pf.$$

Hence the intrinsic energy of any part of the field, as existing in the form of electric displacement, is

$$\tfrac{1}{2}\Sigma(Pf+Qg+Rh)dV.$$

The total energy existing in the field is therefore

$$E=\Sigma\left\{\frac{1}{8\pi}(\alpha\mu\alpha+\beta\mu\beta+\gamma\mu\gamma)+\tfrac{1}{2}(Pf+Qg+Rh)\right\}dV \qquad \ldots\ldots(\mathrm{I}).$$

The first term of this expression depends on the magnetization of the field, and is explained on our theory by actual motion of some kind. The second term depends on the electric polarization of the field, and is explained on our theory by strain of some kind in an elastic medium.

(73) I have on a former occasion* attempted to describe a particular kind of motion and a particular kind of strain, so arranged as to account for the phenomena. In the present paper I avoid any hypothesis of this kind; and in using such words as electric momentum and electric elasticity in reference to the known phenomena of the induction of currents and the polarization of dielectrics, I wish merely to direct the mind of the reader to mechanical phenomena which will assist him in understanding the electrical ones. All such phrases in the present paper are to be considered as illustrative, not as explanatory.

(74) In speaking of the Energy of the field, however, I wish to be understood literally. All energy is the same as mechanical energy, whether it exists in the form of motion or in that of elasticity, or in any other form. The energy in electromagnetic phenomena is mechanical energy. The only question is, Where does it reside? On the old theories it resides in the electrified bodies, conducting circuits, and magnets, in the form of an unknown quality called potential energy, or the power of producing certain effects at a distance. On our theory it resides in the electromagnetic field, in the space surrounding the electrified and magnetic bodies, as well as in those bodies themselves, and is in two different forms, which may be described without hypothesis as magnetic polarization and electric polarization, or according to a very probable hypothesis, as the motion and the strain of one and the same medium.

(75) The conclusions arrived at in the present paper are independent of this hypothesis, being deduced from experimental facts of three kinds:

1. The induction of electric currents by the increase or diminution of neighbouring currents according to the changes in the lines of force passing through the circuit.

* "On Physical Lines of Force," *Philosophical Magazine*, 1861–62.

2. The distribution of magnetic intensity according to the variations of a magnetic potential.

3. The induction (or influence) of statical electricity through dielectrics.

We may now proceed to demonstrate from these principles the existence and laws of the mechanical forces which act upon electric currents, magnets, and electrified bodies placed in the electromagnetic field.

PART IV

# Mechanical Actions in the Field

### Mechanical Force on a Moveable Conductor

(76) We have shewn (§§ 34 & 35) that the work done by the electromagnetic forces in aiding the motion of a conductor is equal to the product of the current in the conductor multiplied by the increment of the electromagnetic momentum due to the motion.

Let a short straight conductor of length $a$ move parallel to itself in the direction of $x$, with its extremities on two parallel conductors. Then the increment of the electromagnetic momentum due to the motion of $a$ will be

$$a\left(\frac{dF}{dx}\frac{dx}{ds}+\frac{dG}{dx}\frac{dy}{ds}+\frac{dH}{dx}\frac{dz}{ds}\right)\delta x.$$

That due to the lengthening of the circuit by increasing the length of the parallel conductors will be

$$-a\left(\frac{dF}{dx}\frac{dx}{ds}+\frac{dF}{dy}\frac{dy}{ds}+\frac{dF}{dz}\frac{dz}{ds}\right)\delta x.$$

The total increment is

$$a\delta x\left\{\frac{dy}{ds}\left(\frac{dG}{dx}-\frac{dF}{dy}\right)-\frac{dz}{ds}\left(\frac{dF}{dz}-\frac{dH}{dx}\right)\right\},$$

which is by the equations of Magnetic Force (B), p. 258,

$$a\delta x\left(\frac{dy}{ds}\mu\gamma-\frac{dz}{ds}\mu\beta\right).$$

Let $X$ be the force acting along the direction of $x$ per unit of length of the conductor, then the work done is $Xa\delta x$.

Let $C$ be the current in the conductor, and let $p'$, $q'$, $r'$ be its components, then

$$X\alpha\delta x = C\alpha\delta x\left(\frac{dy}{ds}\mu\gamma - \frac{dz}{ds}\mu\beta\right),$$

or

Similarly,

$$\left.\begin{aligned} X &= \mu\gamma q' - \mu\beta r' \\ Y &= \mu\alpha r' - \mu\gamma p' \\ Z &= \mu\beta p' - \mu\alpha q' \end{aligned}\right\} \quad \ldots\ldots\ldots\ldots\ldots\ldots (J).$$

These are the equations which determine the mechanical force acting on a conductor carrying a current. The force is perpendicular to the current and to the lines of force, and is measured by the area of the parallelogram formed by lines parallel to the current and lines of force, and proportional to their intensities.

## Mechanical Force on a Magnet

(77) In any part of the field not traversed by electric currents the distribution of magnetic intensity may be represented by the differential coefficients of a function which may be called the magnetic potential. When there are no currents in the field, this quantity has a single value for each point. When there are currents, the potential has a series of values at each point, but its differential coefficients have only one value, namely,

$$\frac{d\phi}{dx} = \alpha, \quad \frac{d\phi}{dy} = \beta, \quad \frac{d\phi}{dz} = \gamma.$$

Substituting these values of $\alpha$, $\beta$, $\gamma$ in the expression (equation 38) for the intrinsic energy of the field, and integrating by parts, it becomes

$$-\Sigma\left\{\phi\frac{1}{8\pi}\left(\frac{d\mu\alpha}{dx} + \frac{d\mu\beta}{dy} + \frac{d\mu\gamma}{dz}\right)\right\} dV.$$

The expression

$$\Sigma\left(\frac{d\mu\alpha}{dx} + \frac{d\mu\beta}{dy} + \frac{d\mu\gamma}{dz}\right) dV = \Sigma m dV \quad \ldots\ldots\ldots (39)$$

indicates the number of lines of magnetic force which have their origin within the space $V$. Now a magnetic pole is known to us only as the origin or termination of lines of magnetic force, and a unit pole is one which has $4\pi$ lines belonging to it, since it produces unit of magnetic intensity at unit of distance over a sphere whose surface is $4\pi$.

Hence if $m$ is the amount of free positive magnetism in unit of volume, the above expression may be written $4\pi m$, and the expression for the energy of the field becomes

$$E = -\Sigma\,(\tfrac{1}{2}\phi m)\,dV \ldots\ldots\ldots\ldots\ldots (40).$$

If there are two magnetic poles $m_1$ and $m_2$ producing potentials $\phi_1$ and $\phi_2$ in the field, then if $m_2$ is moved a distance $dx$, and is urged in that direction by a force $X$, then the work done is $Xdx$, and the decrease of energy in the field is

$$d\{\tfrac{1}{2}\,(\phi_1 + \phi_2)\,(m_1 + m_2)\},$$

and these must be equal by the principle of Conservation of Energy.

Since the distribution $\phi_1$ is determined by $m_1$, and $\phi_2$ by $m_2$, the quantities $\phi_1 m_1$ and $\phi_2 m_2$ will remain constant.

It can be shewn also, as Green has proved (Essay, p. 10), that

$$m_1\phi_2 = m_2\phi_1,$$

so that we get

$$Xdx = d(m_2\phi_1),$$

or

$$X = m_2\frac{d\phi_1}{dx} = m_2\alpha_1,$$

where $\alpha_1$ represents the magnetic intensity due to $m_1$. Similarly, ......(K).

$$Y = m_2\beta_1,$$

$$Z = m_2\gamma_1.$$

So that a magnetic pole is urged in the direction of the lines of magnetic force with a force equal to the product of the strength of the pole and the magnetic intensity.

(78) If a single magnetic pole, that is, one pole of a very long magnet, be placed in the field, the only solution of $\phi$ is

$$\phi_1 = -\frac{m_1}{\mu}\frac{1}{r} \quad \text{..................(41),}$$

where $m_1$ is the strength of the pole, and $r$ the distance from it.

The repulsion between two poles of strength $m_1$ and $m_2$ is

$$m_2 \frac{d\phi_1}{dr} = \frac{m_1 m_2}{\mu r^2} \quad \text{..................(42).}$$

In air or any medium in which $\mu = 1$ this is simply $\frac{m_1 m_2}{r^2}$, but in other media the force acting between two given magnetic poles is inversely proportional to the coefficient of magnetic induction for the medium. This may be explained by the magnetization of the medium induced by the action of the poles.

## Mechanical Force on an Electrified Body

(79) If there is no motion or change of strength of currents or magnets in the field, the electromotive force is entirely due to variation of electric potential, and we shall have (§ 65)

$$P = -\frac{d\Psi}{dx}, \quad Q = -\frac{d\Psi}{dy}, \quad R = -\frac{d\Psi}{dz}.$$

Integrating by parts the expression (I) for the energy due to electric displacement, and remembering that $P$, $Q$, $R$ vanish at an infinite distance, it becomes

$$\tfrac{1}{2}\Sigma \left\{ \Psi \left( \frac{df}{dx} + \frac{dg}{dy} + \frac{dh}{dz} \right) \right\} dV,$$

or by the equation of Free Electricity (G), p. 263,

$$-\tfrac{1}{2}\Sigma\,(\Psi e)\,dV.$$

By the same demonstration as was used in the case of the mechanical action on a magnet, it may be shewn that the mechanical force on a small body containing a quantity $e_2$ of free electricity placed in a field whose potential arising from other electrified bodies is $\Psi_1$, has for components

$$\left.\begin{aligned} X &= e_2 \frac{d\Psi_1}{dx} = -P_1 e_2 \\ Y &= e_2 \frac{d\Psi_1}{dy} = -Q_1 e_2 \\ Z &= e_2 \frac{d\Psi_1}{dz} = -R_1 e_2 \end{aligned}\right\} \quad \text{(D).}$$

So that an electrified body is urged in the direction of the electromotive force with a force equal to the product of the quantity of free electricity and the electromotive force.

If the electrification of the field arises from the presence of a small electrified body containing $e_1$ of free electricity, the only solution of $\Psi_1$ is

$$\Psi_1 = \frac{k}{4\pi}\frac{e_1}{r} \quad \text{(43),}$$

where $r$ is the distance from the electrified body.

The repulsion between two electrified bodies $e_1$, $e_2$ is therefore

$$e_2 \frac{d\Psi_1}{dr} = \frac{k}{4\pi}\frac{e_1 e_2}{r^2} \quad \text{(44).}$$

## Measurement of Electrostatic Effects

(80) The quantities with which we have had to do have been hitherto expressed in terms of the Electromagnetic System of measurement, which is founded on the mechanical action between currents. The electrostatic system of measurement is founded on the mechanical action between electrified bodies, and is independent of, and incompatible with, the electromagnetic system; so that the units of the different kinds of quantity have different values according to the system we adopt, and to pass from the one system to the other, a reduction of all the quantities is required.

According to the electrostatic system, the repulsion between two small bodies charged with quantities $\eta_1$, $\eta_2$ of electricity is

$$\frac{\eta_1\eta_2}{r^2},$$

where $r$ is the distance between them.

Let the relation of the two systems be such that one electromagnetic unit of electricity contains $v$ electrostatic units; then $\eta_1 = ve_1$ and $\eta_2 = ve_2$, and this repulsion becomes

$$v^2\frac{e_1e_2}{r^2} = \frac{k}{4\pi}\frac{e_1e_2}{r^2} \text{ by equation (44)} \quad \ldots\ldots\ldots(45),$$

whence $k$, the coefficient of "electric elasticity" in the medium in which the experiments are made, *i. e.* common air, is related to $v$, the number of electrostatic units in one electromagnetic unit, by the equation

$$k = 4\pi v^2 \quad \ldots\ldots\ldots\ldots\ldots\ldots(46).$$

The quantity $v$ may be determined by experiment in several ways. According to the experiments of MM. Weber and Kohlrausch,

$$v = 310{,}740{,}000 \text{ metres per second.}$$

(81) It appears from this investigation, that if we assume that the medium which constitutes the electromagnetic field is, when dielectric, capable of receiving in every part of it an electric polarization, in which the opposite sides of every element into which we may conceive the medium divided are oppositely electrified, and if we also assume that this polarization or electric displacement is proportional to the electromotive force which produces or maintains it, then we can shew that electrified bodies in a dielectric medium will act on one another with forces obeying the same laws as are established by experiment.

The energy, by the expenditure of which electrical attractions and repulsions are produced, we suppose to be stored up in the dielectric medium which surrounds the electrified bodies, and not on the surface of those bodies themselves, which on our theory are merely the bounding surfaces of the air or other dielectric in which the true springs of action are to be sought.

## Note on the Attraction of Gravitation

(82) After tracing to the action of the surrounding medium both the magnetic and the electric attractions and repulsions, and finding them to depend on the inverse square of the distance, we are naturally led to inquire whether the attraction of gravitation, which follows the same law of the distance, is not also traceable to the action of a surrounding medium.

Gravitation differs from magnetism and electricity in this; that the bodies concerned are all of the same kind, instead of being of opposite signs, like magnetic poles and electrified bodies, and that the force between these bodies is an attraction and not a repulsion, as is the case between like electric and magnetic bodies.

The lines of gravitating force near two dense bodies are exactly of the same form as the lines of magnetic force near two poles of the same name; but whereas the poles are repelled, the bodies are attracted. Let $E$ be the intrinsic energy of the field surrounding two gravitating bodies $M_1$, $M_2$, and let $E'$ be the intrinsic energy of the field surrounding two magnetic poles, $m_1$, $m_2$ equal in numerical value to $M_1$, $M_2$, and let $X$ be the gravitating force acting during the displacement $\delta x$, and $X'$ the magnetic force,

$$X\delta x = \delta E, \qquad X'\delta x = \delta E';$$

now $X$ and $X'$ are equal in numerical value, but of opposite signs; so that

$$\delta E = -\,\delta E',$$

or

$$E = C - E'$$

$$= C - \Sigma \frac{1}{8\pi}(\alpha^2 + \beta^2 + \gamma^2)\,dV,$$

where $\alpha$, $\beta$, $\gamma$ are the components of magnetic intensity. If $R$ be the resultant gravitating force, and $R'$ the resultant magnetic force at a corresponding part of the field,

$$R = -R', \text{ and } \alpha^2 + \beta^2 + \gamma^2 = R^2 = R'^2.$$

Hence

$$E = C - \Sigma \frac{1}{8\pi} R^2 dV \quad \ldots\ldots\ldots\ldots\ldots\ldots\ldots\ldots (47).$$

The intrinsic energy of the field of gravitation must therefore be less wherever there is a resultant gravitating force.

As energy is essentially positive, it is impossible for any part of space to have negative intrinsic energy. Hence those parts of space in which there is no resultant force, such as the points of equilibrium in the space between the different bodies of a system, and within the substance of each body, must have an intrinsic energy per unit of volume greater than

$$\frac{1}{8\pi} R^2,$$

where $R$ is the greatest possible value of the intensity of gravitating force in any part of the universe.

The assumption, therefore, that gravitation arises from the action of the surrounding medium in the way pointed out, leads to the conclusion that every part of this medium possesses, when undisturbed, an enormous intrinsic energy, and that the presence of dense bodies influences the medium so as to diminish this energy wherever there is a resultant atraction.

As I am unable to understand in what way a medium can possess such properties, I cannot go any further in this direction in searching for the cause of gravitation.

[*Part V, which has been omitted, deals with condensers, specific inductive capacity and electric absorption, including residual charge.*]

PART VI

# Electromagnetic Theory of Light

(91) At the commencement of this paper we made use of the optical hypothesis of an elastic medium through which the vibrations of light are propagated, in order to shew that we have warrantable grounds for seeking, in the same medium, the cause of other phenomena as well as those of light. We then examined electromagnetic phenomena, seeking for their explanation in the properties of the field which surrounds the electrified or magnetic bodies. In this way we arrived at certain equations expressing certain properties of the electromagnetic field. We now proceed to investigate whether these properties of that which constitutes the electromagnetic field, deduced from electromagnetic phenomena alone, are sufficient to explain the propagation of light through the same substance.

(92) Let us suppose that a plane wave whose direction cosines are $l$, $m$, $n$ is propagated through the field with a velocity $V$. Then all the electromagnetic functions will be functions of

$$w = lx + my + nz - Vt.$$

The equations of Magnetic Force (B), p. 258, will become

$$\mu\alpha = m\frac{dH}{dw} - n\frac{dG}{dw},$$

$$\mu\beta = n\frac{dF}{dw} - l\frac{dH}{dw},$$

$$\mu\gamma = l\frac{dG}{dw} - m\frac{dF}{dw}.$$

If we multiply these equations respectively by $l$, $m$, $n$, and add, we find

$$l\mu\alpha + m\mu\beta + n\mu\gamma = 0 \quad \ldots\ldots\ldots\ldots\ldots (62),$$

which shews that the direction of the magnetization must be in the plane of the wave.

(93) If we combine the equations of Magnetic Force (B) with those of Electric Currents (C), and put for brevity

$$\frac{dF}{dx} + \frac{dG}{dy} + \frac{dH}{dz} = J, \text{ and } \frac{d^2}{dx^2} + \frac{d^2}{dy^2} + \frac{d^2}{dz^2} = \nabla^2 \quad \ldots (63),$$

$$\left.\begin{aligned} 4\pi\mu p' &= \frac{dJ}{dx} - \nabla^2 F \\ 4\pi\mu q' &= \frac{dJ}{dy} - \nabla^2 G \\ 4\pi\mu r' &= \frac{dJ}{dz} - \nabla^2 H \end{aligned}\right\} \quad \ldots\ldots\ldots\ldots\ldots (64).$$

If the medium in the field is a perfect dielectric there is no true conduction, and the currents $p'$, $q'$, $r'$ are only variations in the electric displacement, or, by the equations of Total Currents (A),

$$p' = \frac{df}{dt}, \qquad q' = \frac{dg}{dt}, \qquad r' = \frac{dh}{dt} \quad \ldots\ldots\ldots\ldots (65).$$

But these electric displacements are caused by electromotive forces, and by the equations of Electric Elasticity (E),

$$P = kf, \qquad Q = kg, \qquad R = kh \quad \ldots\ldots\ldots (66).$$

These electromotive forces are due to the variations either of the electromagnetic or the electrostatic functions, as there is no motion of conductors in the field; so that the equations of electromotive force (D) are

$$\left.\begin{aligned} P &= -\frac{dF}{dt} - \frac{d\Psi}{dx} \\ Q &= -\frac{dG}{dt} - \frac{d\Psi}{dy} \\ R &= -\frac{dH}{dt} - \frac{d\Psi}{dz} \end{aligned}\right\} \quad \ldots\ldots\ldots\ldots\ldots (67).$$

(94) Combining these equations, we obtain the following:—

$$\left.\begin{aligned} k\left(\frac{dJ}{dx} - \nabla^2 F\right) + 4\pi\mu\left(\frac{d^2 F}{dt^2} + \frac{d^2\Psi}{dxdt}\right) = 0 \\ k\left(\frac{dJ}{dy} - \nabla^2 G\right) + 4\pi\mu\left(\frac{d^2 G}{dt^2} + \frac{d^2\Psi}{dydt}\right) = 0 \\ k\left(\frac{dJ}{dz} - \nabla^2 H\right) + 4\pi\mu\left(\frac{d^2 H}{dt^2} + \frac{d^2\Psi}{dzdt}\right) = 0 \end{aligned}\right\} \quad \ldots\ldots(68).$$

If we differentiate the third of these equations with respect to $y$, and the second with respect to $z$, and substract, $J$ and $\Psi$ disappear, and by remembering the equations (B) of magnetic force, the results may be written

$$\left.\begin{aligned} k\,\nabla^2\mu\alpha &= 4\pi\mu\frac{d^2}{dt^2}\mu\alpha \\ k\,\nabla^2\mu\beta &= 4\pi\mu\frac{d^2}{dt^2}\mu\beta \\ k\,\nabla^2\mu\gamma &= 4\pi\mu\frac{d^2}{dt^2}\mu\gamma \end{aligned}\right\} \quad \ldots\ldots\ldots\ldots\ldots(69).$$

(95) If we assume that $\alpha$, $\beta$, $\gamma$ are functions of $lx + my + nz - Vt = w$, the first equation becomes

$$k\mu\frac{d^2\alpha}{dw^2} = 4\pi\mu^2 V^2\frac{d^2\alpha}{dw^2} \quad \ldots\ldots\ldots\ldots\ldots(70),$$

or

$$V = \pm\sqrt{\frac{k}{4\pi\mu}} \quad \ldots\ldots\ldots\ldots\ldots\ldots(71).$$

The other equations give the same value for $V$, so that the wave is propagated in either direction with a velocity $V$.

This wave consists entirely of magnetic disturbances, the direction of magnetization being in the plane of the wave. No magnetic disturbance whose direction of magnetization is not in the plane of the wave can be propagated as a plane wave at all.

Hence magnetic disturbances propagated through the electromagnetic field agree with light in this, that the disturbance at any point is transverse to the direction of propagation, and such waves may have all the properties of polarized light.

(96) The only medium in which experiments have been made to determine the value of $k$ is air, in which $\mu = 1$, and therefore, by equation (46),

$$V = v \quad \ldots\ldots\ldots\ldots\ldots\ldots (72).$$

By the electromagnetic experiments of MM. Weber and Kohlrausch*,

$$v = 310{,}740{,}000 \text{ metres per second}$$

is the number of electrostatic units in one electromagnetic unit of electricity, and this, according to our result, should be equal to the velocity of light in air or vacuum.

The velocity of light in air, by M. Fizeau's† experiments, is

$$V = 314{,}858{,}000;$$

according to the more accurate experiments of M. Foucault‡,

$$V = 298{,}000{,}000.$$

The velocity of light in the space surrounding the earth, deduced from the coefficient of aberration and the received value of the radius of the earth's orbit, is

$$V = 308{,}000{,}000.$$

(97) Hence the velocity of light deduced from experiment agrees sufficiently well with the value of $v$ deduced from the only set of experiments as we yet possess. The value of $v$ was determined by measuring the electromotive force with which a condenser of known capacity was charged, and then discharging the condenser through a galvanometer, so as to measure the quantity of electricity in it in electromagnetic measure. The only use made of light in the experiment was to see the instruments. The value of $V$ found by M. Foucault was obtained by determining the angle through which a revolving mirror turned, while the light reflected

* *Leipzig Transactions*, Vol. v. (1857), p. 260, or Poggendorff's *Annalen*, Aug. 1856, p. 10.

† *Comptes Rendus*, Vol. XXIX. (1849), p. 90.

‡ Ibid. Vol. LV. (1862), pp. 501, 792.

from it went and returned along a measured course. No use whatever was made of electricity or magnetism.

The agreement of the results seems to shew that light and magnetism are affections of the same substance, and that light is an electromagnetic disturbance propagated through the field according to electromagnetic laws.

(98) Let us now go back upon the equations in (94), in which the quantities $J$ and $\Psi$ occur, to see whether any other kind of disturbance can be propagated through the medium depending on these quantities which disappeared from the final equations.

If we determine $\chi$ from the equation

$$\nabla^2\chi = \frac{d^2\chi}{dx^2} + \frac{d^2\chi}{dy^2} + \frac{d^2\chi}{dz^2} = J \quad \ldots\ldots\ldots\ldots (73),$$

and $F'$, $G'$, $H'$ from the equations

$$F' = F - \frac{d\chi}{dx}, \quad G' = G - \frac{d\chi}{dy}, \quad H' = H - \frac{d\chi}{dz} \quad \ldots\ldots (74),$$

then

$$\frac{dF'}{dx} + \frac{dG'}{dy} + \frac{dH'}{dz} = 0 \quad \ldots\ldots\ldots\ldots (75),$$

and the equations in (94) become of the form

$$k\,\nabla^2 F' = 4\pi\mu\left\{\frac{d^2F'}{dt^2} + \frac{d}{dxdt}\left(\Psi + \frac{d\chi}{dt}\right)\right\} \quad \ldots\ldots (76).$$

Differentiating the three equations with respect to $x$, $y$, and $z$, and adding, we find that

$$\Psi = -\frac{d\chi}{dt} + \phi\,(x, y, z) \quad \ldots\ldots\ldots\ldots (77),$$

and that

$$\left.\begin{aligned} k\,\nabla^2 F' &= 4\pi\mu\frac{d^2F'}{dt^2} \\ k\,\nabla^2 G' &= 4\pi\mu\frac{d^2G'}{dt^2} \\ k\,\nabla^2 H' &= 4\pi\mu\frac{d^2H'}{dt^2} \end{aligned}\right\} \quad \ldots\ldots\ldots\ldots (78).$$

Hence the disturbances indicated by $F'$, $G'$, $H'$ are propagated with the velocity $V = \sqrt{\frac{k}{4\pi\mu}}$ through the field; and since

$$\frac{dF'}{dx} + \frac{dG'}{dy} + \frac{dH'}{dz} = 0,$$

the resultant of these disturbances is in the plane of the wave.

(99) The remaining part of the total disturbances $F$, $G$, $H$ being the part depending on $\chi$, is subject to no condition except that expressed in the equation

$$\frac{d\Psi}{dt} + \frac{d^2\chi}{dt^2} = 0.$$

If we perform the operation $\nabla^2$ on this equation, it becomes

$$ke = \frac{dJ}{dt} - k\,\nabla^2\phi\,(x, y, z) \ldots\ldots\ldots\ldots (79).$$

Since the medium is a perfect insulator, $e$, the free electricity, is immoveable, and therefore $\frac{dJ}{dt}$ is a function of $x$, $y$, $z$, and the value of $J$ is either constant or zero, or uniformly increasing or diminishing with the time; so that no disturbance depending on $J$ can be propagated as a wave.

(100) The equations of the electromagnetic field, deduced from purely experimental evidence, shew that transversal vibrations only can be propagated. If we were to go beyond our experimental knowledge and to assign a definite density to a substance which we should call the electric fluid, and select either vitreous or resinous electricity as the representative of that fluid, then we might have normal vibrations propagated with a velocity depending on this density. We have, however, no evidence as to the density of electricity, as we do not even know whether to consider vitreous electricity as a substance or as the absence of a substance.

Hence electromagnetic science leads to exactly the same conclusions as optical science with respect to the direction of the disturbances which can be propagated through the field; both affirm the propagation of transverse vibrations, and both give the same velocity of propagation. On the other hand, both sciences are at a loss when called on to affirm or deny the existence of normal vibrations.

### Relation between the Index of Refraction and the Electromagnetic Character of the substance

(101) The velocity of light in a medium, according to the Undulatory Theory, is

$$\frac{1}{i} V_0,$$

where $i$ is the index of refraction and $V_0$ is the velocity in vacuum. The velocity, according to the Electromagnetic Theory, is

$$\sqrt{\frac{k}{4\pi\mu}},$$

where, by equations (49) and (71), $k = \frac{1}{D} k_0$, and $k_0 = 4\pi V_0^2$.

Hence

$$D = \frac{i^2}{\mu} \quad \text{.....................(80)},$$

or the Specific Inductive Capacity is equal to the square of the index of refraction divided by the coefficient of magnetic induction.

[*We omit paragraphs* (102)–(105) *which deal with the propagation of electromagnetic disturbances in crystalline media.*]

### Relation between Electric Resistance and Transparency

(106) If the medium, instead of being a perfect insulator, is a conductor whose resistance per unit of volume is $\rho$, then there will be not only electric displacements, but true currents of conduction in which electrical energy is transformed into heat, and the

undulation is thereby weakened. To determine the coefficient of absorption, let us investigate the propagation along the axis of $x$ of the transverse disturbance $G$.

By the former equations

$$\frac{d^2G}{dx^2} = -4\pi\mu(q')$$

$$= -4\pi\mu\left(\frac{df}{dt} + q\right) \text{ by (A),}$$

$$\frac{d^2G}{dx^2} = +4\pi\mu\left(\frac{1}{k}\frac{d^2G}{dt^2} - \frac{1}{\rho}\frac{dG}{dt}\right) \text{ by (E) and (F)} \ldots\ldots(95).$$

If $G$ is of the form

$$G = e^{-px}\cos(qx + nt) \ldots\ldots\ldots\ldots(96),$$

we find that

$$p = \frac{2\pi\mu}{\rho}\frac{n}{q} = \frac{2\pi\mu}{\rho}\frac{V}{i} \ldots\ldots\ldots\ldots(97),$$

where $V$ is the velocity of light in air, and $i$ is the index of refraction. The proportion of incident light transmitted through the thickness $x$ is

$$e^{-2px} \ldots\ldots\ldots\ldots\ldots\ldots(98).$$

Let $R$ be the resistance in electromagnetic measure of a plate of the substance whose thickness is $x$, breadth $b$, and length $l$, then

$$R = \frac{l\rho}{bx},$$

$$2px = 4\pi\mu\frac{V}{i}\frac{l}{bR} \ldots\ldots\ldots\ldots(99)$$

(107) Most transparent solid bodies are good insulators, whereas all good conductors are very opaque.

Electrolytes allow a current to pass easily and yet are often very transparent. We may suppose, however, that in the rapidly alternating vibrations of light, the electromotive forces act for so short a time that they are unable to effect a complete separation between the particles in combination, so that when the force is reversed the particles oscillate into their former position without loss of energy.

Gold, silver, and platinum are good conductors, and yet when reduced to sufficiently thin plates they allow light to pass through them. If the resistance of gold is the same for electromotive forces of short period as for those with which we make experiments, the amount of light which passes through a piece of gold-leaf, of which the resistance was determined by Mr C. Hockin, would be only $10^{-50}$ of the incident light, a totally imperceptible quantity. I find that between $\frac{1}{500}$ and $\frac{1}{1000}$ of green light gets through such gold-leaf. Much of this is transmitted through holes and cracks; there is enough, however, transmitted through the gold itself to give a strong green hue to the transmitted light. This result cannot be reconciled with the electromagnetic theory of light, unless we suppose that there is less loss of energy when the electromotive forces are reversed with the rapidity of the vibrations of light than when they act for sensible times, as in our experiments.

### Absolute Values of the Electromotive and Magnetic Forces called into play in the Propagation of Light

(108) If the equation of propagation of light is

$$F = A \cos \frac{2\pi}{\lambda}(z - Vt),$$

the electromotive force will be

$$P = -A\frac{2\pi}{\lambda} V \sin \frac{2\pi}{\lambda}(z - Vt);$$

and the energy per unit of volume will be

$$\frac{P^2}{8\pi\mu V^2},$$

where $P$ represents the greatest value of the electromotive force. Half of this consists of magnetic and half of electric energy.

The energy passing through a unit of area is

$$W = \frac{P^2}{8\pi\mu V};$$

so that

$$P = \sqrt{8\pi\mu VW},$$

where $V$ is the velocity of light, and $W$ is the energy communicated to unit of area by the light of a second

According to Pouillet's data, as calculated by Professor W. Thomson*, the mechanical value of direct sunlight at the Earth is

83·4 foot-pounds per second per square foot.

This gives the maximum value of $P$ in direct sunlight at the Earth's distance from the Sun,

$P = 60{,}000{,}000$ [see *Note* below]

or about 600 Daniell's cells per metre.

At the Sun's surface the value of $P$ would be about

13,000 Daniell's cells per metre.

At the Earth the maximum magnetic force would be 0·193†.

At the Sun it would be 4·13.

These electromotive and magnetic forces must be conceived to be reversed twice in every vibration of light; that is, more than a thousand million million times in a second.

*Note on Maxwell's figures for sunlight.* Something seems to have gone wrong with these calculations. The figures for the earth's surface were repeated in the *Treatise*, and in his notes to the third edition J. J. Thomson comments as follows on the values given for these intensities: "I have not been able to verify these numbers, if we assume $v = 3 \times 10^{10}$, the mean energy in one cc. of sunlight is, according to Pouillet's data, as quoted by Thomson, $3\cdot92\ 10^{-5}$ ergs, the corresponding values of $P$ and $\beta$ as given by (24) are in c.g.s. units,

$P = 9\cdot42 \times 10^8$ or 9·42 volts per centimetre.‡

$\beta = 0\cdot0314$ or rather more than a sixth of the earth's horizontal magnetic force."‡

* *Transactions of the Royal Society of Edinburgh*, 1854 ("Mechanical Energies of the Solar System").

† The horizontal magnetic force at Kew is about 1·76 in metrical units.

‡ Modern values would give $P = 10\cdot2$ volts/cm; $\beta = 0\cdot0336$.

The figures for the sun's surface also appear to be incorrect. The sun's radius is 432,000 miles and the distance of the earth from the sun is about 93,000,000 miles. The vectors in the light will vary inversely as the distance, so that at the sun's surface their values will be 215 times what they are at the earth. Maxwell's ratio is 21·6 and is thus out by a factor of 10. Correcting both errors the values at the sun's surface become,

$P = 2{,}024$ volts per centimetre,

or about 200,000 Daniell's cells per metre.

$\beta = 6 \cdot 75$ gauss.

[*Part VII of this paper is headed "Calculation of the Coefficients of Electromagnetic Induction". Lack of space has necessitated its omission.*]

# Index

Abbott, B. 14, 17
Action at a distance 8, 9
Adams, W. G. 98
Airy, G. 89
Ampère, A. M. 3, 5, 6, 15, 101
Ampere, the determination of the 42
Arago 22, 31
Athenaeum Club 19
Austral pole 26

Babinet 22
Bacon 4
Ball bearings 114f.
Barlow, J. 89
Barnhard, George 21
Barnhard, Sarah 17
Bence, Jones 12, 18, 33
Berzelius 11, 12
Biot 3, 22
Bismuth, magnecrystallic axis of 35
Boreal pole 26
Boscovitch 9, 10, 75
Boyle 4
Brewster, D. 93

Caloric 4
Cambridge Philosophical Society 12, 102
Campbell 94
Carlisle 10
Cavendish, Hon. H. 77, 99
Cavendish Laboratory 99, 100
Cay, J. 93
Chemical combination 11
Chemical History of a Candle 22
Civil List Pension 20
Clausius 4
Clement 15
Clerk Maxwell, Frances 92
Clerk Maxwell, John 92
Col de Tende 15
Common electricity 26
Conservation of energy 6
Contact theory 8
Corpuscular theory of light 5
Coulomb 8
Courtois 15
Cruickshank 10
Current electricity 5, 66, 67, 82
Cuvier 15

Davy, Lady 17
Davy, Sir H. 6, 10, 11, 14
De la Rive, Auguste 11, 16, 17
De la Rive, Prof. G. 11, 16, 17
Desormes 15
Dewar, Frances Mary 96
Diamagnetism 35, 84, 85, 124
Diamonds, burning, in oxygen 15
Displacement current 107f., 120f.
Dumas 22
Dynamo 31

Edinburgh Academy 92
Edinburgh Society of Arts 93
Electric current 5
Electric current, Faraday's theory of 82
Electric fluids 4, 5
Electricity, common 26
Electricity of tension 26
Electrolysis 10, 11, 66, 77, 78
Electromagnetic induction 24f.
Electromotor 26
Electroscopic force 8

Electro-tonic state 26–31, 82, 83, 105, 107
Energy of magnetic field 50f.
Epinus 77
Euler 9

Fechner 68
Fizeau 5
Flux cutting 46
Flux theory 103
Forbes 93, 96
Foucault 5
Fourier 7

Garnett 94
Gay-Lussac 15
Glenlair 92, 93, 95, 96
Gravity 26, 274

Hall effect 66
Hampton Court 23
Hare, Prof. 79
Haswell Colliery 22
Heat 4
Heaviside, O. 24, 128
Helmholtz 3, 4, 48, 49, 50, 54, 55, 62
Henry 3
Hertz 130
Hooke 4
Humboldt 15
Huxtable 14, 19

Imponderable matter 4, 6, 80

Jenkin, W. 33
Joule 7, 48

Kelvin, Lord 3, 4, 9, 25, 50, 59, 93, 98, 99

Laplace 3, 9
Lines of force 6, 7, 24, 25, 29, 34–39, 46, 62, 75, 76, 81, 89, 90, 97
Lorenz 44, 71, 72

Macquerier 13
Magnecrystallic axis 35
Magnetic fluids 5
Magnetism, Ampère's theory of 6
Marischal College, Aberdeen 95
Marked pole 26
McCunn 98
Mechanical equivalent of heat 7, 48, 54, 72
Melbourne, Lord 20
Mercurial poisoning 21
Molecular currents 6

Neumann 3, 46–48, 68
Newton 4
Nichol 93
Nicholson 10

Oersted 3, 5, 25, 74
Ohm 7, 8, 34
Oxygen, paramagnetism of 86

Paramagnetism 23, 35, 86, 88, 124
Peel, Sir R. 20
Pension, Civil List 20
Pergrinus 6
Phillips 27, 31
Physical lines of force 107, 110
Poisson 8, 9, 77
Polarization, electric 5, 77–79, 82
Polarized light, rotation of plane of 124
Pomeroy 98
Ponderable matter 4, 6, 80
Priestley 8

Relativity 69
Retarded potentials 9
Ribeau, G. 13

Robison 8
Roche, de la 14
Routh 94
Royal Institution 14, 17, 19, 20, 22
Royal Society, Faraday's election 18
Royal Society of Edinburgh 93
Rumford 4

Sandemanian Church 17, 18
Saturn rings 98
Savart 3
Seebeck 7
Self-induction 28, 33, 34
Space 4, 35, 36, 76, 86–88

Tait, P. G. 93, 94, 96
Tatum 13
Taylor R., 75
Thomson, W., and Lord Kelvin 3, 9, 25, 50, 59, 93, 98, 99
*Treatise on Electricity and Magnetism* 98, 99, 101, 102
*Treatise on Heat* 99
Trinity House 20, 23
Tyndall 74

Vacuum, ideas of 4, 35, 36, 76, 86–88
Vector potential 47, 105
Volt, determination of 42
Volta 16
Vortices, theory of 108, 110f.

Wave theory of light 5
Weber 3, 5, 63f.
Wedderburn, Mrs. 92
Wollaston 18
Woolwich, Faraday's appointment to 19

Young 4

Printed in Dunstable, United Kingdom

91343792R00178